Breakthrough

Breakthrough

Stories and Strategies of Radical Innovation

Mark Stefik and Barbara Stefik

The MIT Press
Cambridge, Massachusetts
London, England

First MIT Press paperback edition, 2006

MIT Press books may be purchased at special quantity discounts for business or sales promotional use. For information, please email special_sales@mitpress.mit .edu or write to Special Sales Department, The MIT Press, 55 Hayward Street, Cambridge, MA 02142.

Set in Sabon by SNP Best-set Typesetter Ltd., Hong Kong. Printed and bound in the United States of America.

Library of Congress Cataloging-in-Publication Data

Stefik, Mark.
Breakthrough : stories and strategies of radical innovation / Mark Stefik and Barbara Stefik.
p. cm.
Includes bibliographical references and index.
ISBN 0-262-19514-3 (alk. paper), 0-262-69337-2 (pb)
1. Technological innovations. 2. Inventions. I. Stefik, Barbara. II. Title.

T173.8.S745 2004
608—dc22   2004042642

10 9 8 7 6 5 4 3

Dedicated to the many brilliant and creative people who have worked at the Palo Alto Research Center, for living a culture of innovation that inspired this book. With gratitude to George Pake, the founder of PARC, who passed away on March 4, 2004. Pake knew how to set direction with incentives and also how to encourage creativity by staying "out of the way." He set an example for research leaders everywhere.

Contents

Preface

Innovation is subtle, complex, and full of surprises. It depends on organizational culture and practice as much as on individual brilliance. This book is for people who are interested in innovation and especially in breakthroughs. It explains how leading companies succeed repeatedly in inventing the future.

Open innovation refers to strategies that companies use to acquire technologies from other companies and to market their own technologies. Open innovation makes good economic sense. However, it is only half of the story. Nobody ever created a breakthrough with open innovation. *Open invention* refers to strategies for creating inventions and breakthroughs. The challenges of the new century require strategies for both open innovation and open invention. Without a culture of innovation that honors the pursuit of breakthroughs, the tendency is to optimize the routine, to pursue incremental improvements, and to resist the truly innovative.

This book is a road map to the ways of invention and innovation. It introduces the terminology and the practices of leading inventors, entrepreneurs, and managers in innovative organizations. Among its goals are to illuminate how innovation works and how breakthroughs are created.

The chapters in part I introduce two questions that must be answered for every successful innovation: What is possible? What is needed?

The chapters in part II focus on invention and on the experiences of creative researchers and inventors at work. Without invention, there can be no innovation. These chapters address the following questions: Do all inventors work the same way? Do researchers work alone, or in teams? What makes breakthrough inventions different from incremental product improvements? What does an "Aha!" feel like? How do repeat inventors foster the conditions under which the best ideas arise? What educational practices foster invention and innovation?

Part III focuses on organizations and on the fostering of innovation. Innovation involves more than lone inventors and engineering teams. The full story of innovation takes place in the innovation ecology, which includes inventors, entrepreneurs, research managers, venture capitalists, universities, and government funding managers. The chapters in this part address the challenges and opportunities for innovation in the new century, asking these questions: What makes radical innovation so difficult? What matters in creating and managing a great research group? In managing research, why do some institutions use a patron model and others a client model? How are the scientific and business practices for innovation changing, and what is their future? How will business strategies for innovation change in the new century? What are the emerging strategies for both open innovation and open invention?

Our approach was to ask people for their stories. We interviewed inventors and others directly engaged in the innovation ecology. We focused on *repeat* inventors and managers rather than on individuals who had only one invention or who had stopped inventing. About half of the stories come from PARC (formerly Xerox PARC). The stories— about research cultures, business cycles, obstacles to radical innovation, and so on—are organized thematically.

We brought different perspectives to the writing. Mark Stefik is an inventor and a research fellow at PARC, where he directs the Information Sciences and Technologies Laboratory. He brought the questions and perspectives of research. Barbara Stefik has a doctorate in transpersonal psychology and is in private practice. She brought a questioning mind, an awareness of the mystery in creativity, and a practice of delving for the essence of experiences.

Acknowledgments

Many thanks to the inventors, the researchers, the research managers, and the business development professionals who shared their stories with us: Maribeth Back, Henry Baird, Ben Bederson, Daniel Bobrow, John Seely Brown, Stuart Card, David Fork, Rich Gold, David Goldberg, John Hennessy, Jerry Hobbs, Bob Kahn, Ron Kaplan, Nobutoshi Kihara, Joshua Lederberg, Larry Leifer, Jock Mackinlay, Pattie Maes, John Riedl, Dave Robson, Daniel Russell, Paul Saffo, Ted Selker, Nick Sheridon, Ben Shneiderman, Diana Smetters, Bettie Steiger, Bob Taylor, Chuck Thacker, and Mark Yim.

Special thanks to Steve Smoliar. His penetrating criticisms and suggestions over several drafts of the book uncovered many issues that we needed to understand. Special thanks also to Ed Feigenbaum, whose comments at the end of the third draft properly sent us back to the drawing board to clarify and bring out the main story. Special thanks to John Seely Brown. He helped us to see more clearly some familiar things that had become invisible, and also how to connect some of the dots in our story. Thanks to all of our colleagues who read early drafts of this book and offered suggestions of all sorts: Ben Bederson, Mark Bernstein, Dana Bloomberg, Daniel Bobrow, Ross Bringans, Steve Cousins, Jim Gibbons, Meg Graham, David Hecht, Johan de Kleer, Rich Domingo, Bill Janssen, Lauri Karttunen, Teresa Lunt, Bob Prior, Dave Robson, Dan Russell, and Paige Turner. Thanks to James Fadiman, who advised us to focus on stories. For many of our reviewers, the subject of innovation and breakthroughs is dear to them and research is a passion. Apologies to any helpful colleagues we forgot to mention. We thank them for their generosity and thoughtfulness.

Thank you to Maia Pindar and Kathy Jarvis of PARC's Information Center for quickly finding information that we requested, and also to Giuliana Lavendel, who is an insider's information resource. Thanks to Brian Tramontana and Mimi Gardner for locating photographs of PARC inventions. Thanks to Dan Murphy and Mark Wong of PARC Creative Services for creating illustrations for the book. Thank you to Sony Corporation for facilitating communication with Nobutoshi Kihara, and for providing translation assistance. Thanks to Eric Stefik, who sent us materials and struck gold for us several times.

As authors we have learned a great deal from interactions with our colleagues both at PARC and elsewhere. Creativity and innovation are vast topics, and our understanding keeps deepening. We have tried to capture the best of our understanding in the stories and reflections in the chapters. Ultimately, the conclusions in this book are those of the authors. They are neither a consensus opinion of the individuals we interviewed nor an official position of the Palo Alto Research Center.

The writing of this book took about three years of weekends and vacations. We took several short trips to quiet places to write and get away from our daily responsibilities and routines. Thank you to the people at Asilomar Conference Center in Pacific Grove, the Quaker retreat center in Ben Lomond, the Calaveras Big Trees campgrounds in Arnold, the Apple Farm in San Luis Obispo, Huddart County Park in San Mateo County, Café Borrone in Menlo Park, and the beach at the end of Kelly Avenue in Half Moon Bay. Everyone accommodated us generously, providing the creative ambiance or supporting the silence and even finding places to charge the batteries for our laptop computers.

Mark Stefik thanks his parents for providing formative early experiences. Beyond trips to science museums in New York City, he remembers Edison's laboratories in New Jersey which gave him a first impression of both invention and an invention company. He remembers the 1962 World's Fair in Seattle with the Space Needle, the monorail, General Electric's Pavilion and its Home of the Future, and the Bell Systems Pavilion with its videophones. He remembers visiting Pittsburgh and seeing the Alcoa building and other aluminum-covered buildings glowing in the public square, and visiting Corning, where the flawed

200-inch mirror made for Mount Palomar was on display. In writing this book he realized how these early experiences and Benson Polytechnic High School in Portland planted seeds for a career in research and technology.

Thanks also to our family—Ryan, Morgan, Nick, and Paige—for keeping things going and for taking care of our dog, Shadow, when we took off to work on the book.

I
Invention and Innovation

1

The Breakthrough Dilemmas

Invention is a flower. Innovation is a weed.
—Bob Metcalfe

Breakthroughs

Breakthroughs take people by surprise. They are rare events, arising from scientific or engineering insights. They are called "breakthroughs" because they do something that most people did not realize was possible. Breakthroughs create something new or satisfy a previously undiscovered need. Big breakthroughs often have uses and effects far beyond what their inventors had in mind. Breakthroughs can launch new industries or transform existing ones.

Few companies actively look for breakthroughs. Interest in breakthroughs is sporadic, tending to arise when companies see trouble ahead because the market demand or profitability of their products is declining. Eventually this decline happens for all technologies. The intensity of interest in funding the kind of research that creates breakthroughs varies. It often correlates with where a company's products are on the technology-adoption curve.

Technology-Adoption Curves

High-technology businesses go through recognizable business cycles. After fragile and delicate beginnings, they grow rapidly and robustly. When they reach maturity they stop growing. Eventually they decline.

Paul Saffo is a senior researcher at the Institute of the Future.[1] When Saffo looks at companies and technology trends, he thinks about the long view. He is familiar with S-curves. For Saffo, technology-adoption curves

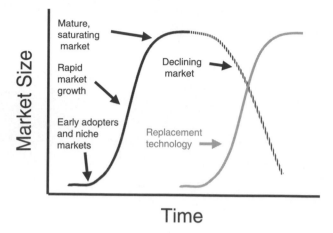

Figure 1.1
The adoption of technologies follows an S-shaped curve. The second curve represents the rise of a new technology that will replace the first technology in the market.

are part of the predictable landscape. They explain deep truths about how the innovation ecology works. Saffo:[2]

There is a pattern here where science makes progress and creates a new technology. The technology starts to grow up and impact people's lives. Eventually it plateaus and the main part becomes routine. Then cost-cutting measures drive manufacturing overseas.

You can plot this out in S-curves. The first phase is the solo inventor, like Doug Engelbart[3] working decades ahead of everyone else. When you start to get closer to the inflection point, you have teams working together, like Apple and its Macintosh team. As the S-curve matures and heads toward the top, the business becomes bigger and more bureaucratic. That's where "creative destruction"[4] comes in because the more something becomes bureaucratized, the more room it leaves at the bottom for individuals and small teams of heretics to redefine the game in new ways. The cost of innovation goes up when a technology becomes bureaucratized.

The *technology-adoption S-curve*[5] (figure 1.1) begins with an invention. In the early part of the curve when the technology is new and immature, the market is small. It is limited to a few early adopters and small niche markets. As a technology matures and reaches the mainstream, businesses enter a phase of rapid growth. They focus on competition with other businesses, improve their products to meet evolving customer needs, and increase their market share. At the market peak, there is often consolidation for market share and economies of scale. When a market saturates, businesses look for other growth opportunities.

The videocassette recorder business and the videotape rental business illustrate the growth and decline of a technology. Video tape recorders were invented in the 1950s, but the first videocassette recorders intended for home use were developed by Sony in 1964.[6] In the early stages, there were legal challenges concerning copyright issues, and there was a standardization struggle between two formats (Beta and VHS). In the 1980s, the rental market grew and many rental stores were opened. Eventually the rental market consolidated; Blockbuster's share was about 40 percent. In the 1990s the VCR business began to decline as a result of competition from other technologies, including cable and satellite television. After DVDs came along, network-enabled DVD mailing services (e.g., Netflix[7]) appeared; by 2002 they had reached about 600,000 customers. This pattern of a slow beginning, a period of rapid growth, a market peaking and consolidating, and competition from new technologies is typical of high-technology industries.

Sometimes companies increase their markets by expanding into new geographic areas. When it is possible, they move manufacturing or other parts of the business overseas, chasing cheaper labor. Eventually, the original business is displaced by new businesses that then enter their own periods of rapid growth, eroding the market of the original technology.

Several factors drive S-curves. During the course of an S-curve, markets grow, companies become bloated, and technical knowledge spreads. Several predictable side effects arise and shape business decisions. Competition begins, markets saturate, manufacturing costs are driven down, margins evaporate, and so on. In his book *Only the Paranoid Survive*, Intel founder Andy Grove put it this way:

Business success contains the seeds of its own destruction. The more successful you are, the more people want a chunk of your business and then another chunk and then another until there is nothing left. . . . A strategic inflection point is a time in the life of a business when its fundamentals are about to change. That change can mean an opportunity to rise to new heights. But it may just as likely signal the beginning of the end.

Another factor—the same one that gives rise to technologies in the first place—is that new ideas arise. Old technologies get displaced by new ones. To thrive, businesses need a path to renewal. Paul Saffo continues:

What happens is that investors start to ask where a dollar invested yields the higher return. Do you invest at the top of an S-curve, trying to hold the old S-curve up at the top? Or do you begin investing in the bottom of the next S-curves to ride up with the new technologies? In that sense the bureaucratization of a maturing field sets the stage for the next wave of innovation.

In this long view of innovation cycles, technologies mature and leave more room at the bottom. *No company or country can expect to dominate a particular technology indefinitely.* Technologies either move away to other regions or are superseded by the next new thing. What creates a new opportunity at the bottom is the investment in the next round of inventions.

Technology-adoption curves reveal the business conditions that shape how a company perceives its need for innovation. Often companies don't see ahead to the erosion of their markets. Sometimes there is a substantial delay between when a downturn begins and when a company recognizes it and acts. To buffer across changing conditions, some companies diversify into multiple kinds of product. A product at its revenue peak can be a "cash cow" for products that are at other points in the cycle. Managed in this way, companies can average out the boom-bust cycle of individual technologies or products. This is the pattern for consumer electronics companies, which create new devices and media every few years.

Although S-curves help to explain what is happening when products and technologies grow and then decline, they are not themselves "the problem." They are a natural part of the cycle of renewal for technologies and innovation. They lead to surprises and crises only when companies are not paying attention.

Trouble at the Top

Today, many products in the information and communications sectors are peaking at the same time. Computers, phones, video games, video projectors, digital cameras, and printers are all based on integrated circuits and packaging technologies. These products are affected in the same way by declining profits and outsourced manufacturing, because the underlying sciences and technologies—semiconductors, computation, modular manufacturing—are essentially the same.[8] These technologies aren't going away, but their period of high profitability and rapid growth seems to be over.

Meanwhile, in other areas of manufacturing beyond semiconductors, increased outsourcing is making products from different companies more alike, since they are increasingly built from the same generic parts in shared factories. With simultaneous maturing, globalization, and displacement, a lot of change is happening at once.

The Innovation Ecology

The future is invented not so much by a heroic loner or by a single company with a great product as by the capacity to combine science, imagination, and business. An *innovation ecology* includes education, research organizations, government funding agencies, technology companies, investors, and consumers. A society's capacity for innovation depends on its innovation ecology. (The information-technology sector provides a powerful example. Roughly following Moore's Law,[9] computers doubled in power, speed, and capacity every 18 months for more than 30 years.) Today, genomics and proteomics are poised for a big run of discoveries. Exactly what discoveries, inventions, and innovations will matter most in the next decades is not yet known, but it is likely that a lot of things will happen. Ambitious scientists and engineers are drawn to the chase. They want to be the ones who make the big discoveries or invent the next big thing. Succeeding in the chase, however, requires expensive equipment, first-rate colleagues, and sustained support. Historically, this combination of resources has been found mainly at corporate research institutions.

The Breakthrough Zone
The scientific knowledge base for innovation is created by basic research, largely at universities, but new technologies are created mainly at corporate research centers. For example, after years of basic research in materials and electronics, the transistor was created at Bell Labs. Bell Labs also created many kinds of communication technologies. Laboratories at Corning Glassworks developed fiber optics after basic research on optics and glass had been done in the academic community for many decades. Laboratories at Texas Instruments and Fairchild Semiconductor invented the first integrated circuits, establishing a scalable technology for modern computers.[10] IBM research created several generations

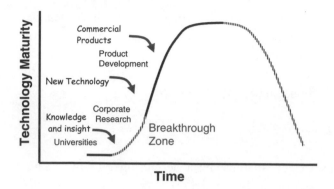

Figure 1.2
The breakthrough zone is the region between basic research and the development of a new and usable technology.

of magnetic storage technology and practical databases. Xerox PARC (Palo Alto Research Center) created the personal computer, the Ethernet, and the laser printer.

The path to new technology has been similar in the biotechnology sector. Restriction enzymes were discovered and methods for sequencing DNA were developed at universities in the United States and the United Kingdom.[11] Subsequently, sequencing methods were refined in industrial labs, and automated DNA-sequencing machines were built. Polymerase chain reaction for amplifying or replicating DNA was discovered at Cetus.[12] In the arena of drug discovery, basic research is followed by systematic searches for effective drugs. With improved technology, these searches are getting faster, but drugs and medical appliances still take years to develop and then several more years to test in clinical trials. Because of the many unknowns and hurdles in developing biotechnology,[13] success usuallly requires genuine breakthroughs.

Basic research, which establishes a knowledge base, takes place largely at universities. However, universities can carry research only so far. Their educational agenda takes precedence, and they cannot focus sustained resources to develop a technology much beyond basic research. They rely heavily on graduate students to carry out the research. Just as the students really master their area, they graduate. Many of them go on to work in corporate research laboratories. These laboratories have historically been the main institutions with the skills and staying power to take a new technology to the point where it can be applied to create products. These labs create teams of the brightest people from each generation to tackle hard problems. This is the center of the breakthrough zone.

A Phase Model for Innovation

Studying 100 years of development in the chemical and aluminum industries, Margaret Graham and Bettye Pruitt[14] characterize innovation in an industry according to five detailed phases. The first is the *borrowing* phase, in which companies borrow the initial knowledge and practices from universities. In the *internalization* phase, companies hire scientists and engineers from the universities and invest in research, hoping to gain advantages in know-how and intellectual property for products they will develop later. In the third phase, characterized as *institutionalization*, an industry sets up its own central research laboratory. As an industry matures, its researchers no longer depend as much on external sources of knowledge; this is the *specialization* phase. In the fifth phase, the technology is mature and companies focus on improving their current products. This phase is called *routinization* because the results become predictable and incremental.

Sometimes, when the market for a technology saturates, a technology company will recognize a need for breakthroughs. To go to the next level, companies must enter a *renewal* phase, essentially repeating earlier parts of the cycle. They can either reinvigorate internal research or go outside to universities or other sources. If a company fails to renew its business, the most likely scenario is for it to retrench in outsourcing manufacturing and routine client-oriented innovation for the ride down from the peak.

Figure 1.3 overlays Graham and Pruitt's phases of R&D development on the technology-adoption S-curve. It shows how the center of gravity for innovation on a technology shifts over time. Innovation typically begins at universities with basic research that yields discoveries and insights. It then moves to corporate research labs, where a usable technology is created. Finally it moves to product-development organizations.

Basic research—the first stage of innovation—takes place over many years. Since much of it is done at universities, it depends for support on government agencies and on corporate sponsors. The next phases of

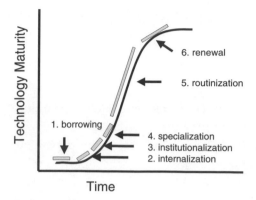

Figure 1.3
Phases in a company's R&D development overlaid on the technology-adoption S-curve.

innovation, which take place in industrial research laboratories and later in product-development organizations, depend largely on private investment by individual companies. This part of the innovation process is vulnerable to instabilities in funding as the fortunes of companies fluctuate and as investment priorities shift.

Losing Sight of Breakthrough Research

Although investment in research and development (R&D) rose in the period 1980–2000, the focus has shifted toward development and toward the short term.[15] Inattention to the long term has resulted in a decline in the kinds of investment that yielded many breakthroughs in the last century.

In the private sector, the demand for short-term profits and quarterly reporting of revenues diverted attention away from long-term perspectives and led to economic churning. In the 1980s, the churning took the form of corporate mergers and acquisitions. Rather than invest in the creation of new technologies, companies sought to acquire technologies and market share by buying other companies. The 1990s saw the rise and fall of the "new economy" and of startup companies. Again, more attention was paid to what could be had quickly than to breakthroughs and sustainable advantages. The focus shifted from long-term research to short-term research and incremental improvements. In the public sector, national laboratories such as Sandia, Lawrence Livermore, and

Oak Ridge began shrinking and redefining their missions. In the private sector, research institutions shrank and some were broken up.[16] In addition, research was redirected across the information-technology sector. To some extent, these changes reflect a healthy shifting of resources away from older technologies and toward fields with greater opportunities for growth, such as the life sciences and biotechnology. Nonetheless, the overall ecology has shifted, and important factors that contributed to success are now weaker than before.

Can other elements of the innovation ecology pick up the slack for corporate research? One way to gain perspective on this question is from the vantage points of institutions at the two ends—the basic research end at universities and the product-application end at startups or development organizations.

John Hennessy has been involved both in business and in academia. He co-founded MIPS Computer Systems (now MIPS Technologies, Inc.). Before becoming president of Stanford, he was the university's provost, its dean of engineering, and a professor of electrical engineering. Hennessy believes that long-term research is fundamentally in the public interest, and that it is becoming difficult for corporations to sponsor such research[17]:

Universities need to do things in the innovation process that are beyond what industry can look at. That role is likely to become larger with either the collapse or redirecting of the large-scale industrial research labs. If you look around at labs like Xerox PARC, Bell Labs, and IBM, these are all places which had 10- or 20-year visions of where they were trying to go. But in the modern corporate environment you simply can't support that kind of work. Basic research is mainly for the public good. It has become too hard to support it in modern corporate America. Nobody wants to do the basic research that will ensure that the industry continues to innovate as opposed to giving them a technology advantage in their next five-year time frame.

One of the things that was absolutely admirable about what happened in industry was the ability to make a large-scale long-term commitment to pursuing a concept. You see this whether you look at Xerox PARC, Bell Labs, or IBM research and you look at innovations like the Alto [figure 1.4] or the transistor or magnetic disks. It is hard to get that kind of collective excitement and focus in a university. It is very difficult to get a commitment that will last for that many years and involve enough people to make that kind of collective breakthrough.

Universities are good at doing discontinuous research, but I don't know whether they can do it on that kind of scale. It depends on how we set ourselves up. We depend on governmental research grants, but those have gotten more focused and short-term over time and are less open ended than they were.

Figure 1.4
An Alto personal computer, developed at PARC in the 1970s. Courtesy of Palo
Alto Research Center.

Many of the people in industry realize that it is necessary to have speculative
long-term research. I think that they would rather pay part of the bill at the
university and then share in the benefits than to pay the whole bill themselves
and have everybody share in the benefits anyway.

Still, I worry about whether we can get the critical mass and scale in universities that is necessary to attack very-large-scale problems. That will be depen-
dent on funding models that bring together creative and interested
philanthropists, government, and industry.

Hennessy's analysis frames a significant challenge for the innovation
ecology. Basic research is in the public interest and will continue at universities. Corporations are increasingly focused on product development
and very-short-term research. The bridge between basic research and
applications—the crucial work of the breakthrough zone—is disappear-
ing. This leaves an innovation gap and raises concern about the health
of the innovation ecology.

During the dotcom craze of the 1990s, it seemed that startups could
do everything. Chuck Thacker is a prolific inventor perhaps best known
for his role in creating the first personal computers. Looking out at
the dotcom phenomenon from his position at the Digital Equipment

Corporation's Systems Research Center, Thacker was worried. Smart people were leaving research laboratories and universities to start companies. Of course many of the companies failed quickly, but new ones were formed just as quickly and some people seemed to be doing very well.[18] "At least until very recently," Thacker recalled, "if you had a good idea you would take it up to Sand Hill Road.[19] I was actually worried until the huge technology collapse that industrial research labs were doomed because of the things that were being done by startups. That is less true now, certainly." Thacker's comments reflect the widespread confusion and uncertainty in the research community about startups during the 1990s. (Weren't the startups doing high technology and everything that the research community did? Did they have a better chance than research labs or universities of getting their ideas out?)

Corporate research labs enjoyed sustained funding that allowed them to pursue problems, whereas universities and startup companies simply couldn't afford to support research groups for long periods.

The Breakthrough Dilemmas

A technological society depends on breakthroughs. The future always brings surprises. The challenges arise from such factors as growing populations, increasingly scarce resources, and a straining ecology. More sophisticated knowledge and technologies are our best hopes for coping with new challenges.

One might expect business, education, and government institutions to be well prepared to the breakthroughs that society will need. As it turns out, however, breakthroughs are quite rare, and creating them is a subtle affair. Businesses, educational institutions, and government funding agencies are all involved in breakthroughs, but most of their attention is necessarily devoted to routine business rather than breakthroughs. By optimizing their routine activities, these institutions can create barriers that get in the way of creating breakthroughs. We call these barriers *the breakthrough dilemmas*.[20]

The Breakthrough Dilemmas for Corporations

Corporations face two breakthrough dilemmas. One arises when corporations stop funding breakthrough research. The other arises when they sustain research funding but fail to exploit the innovations. In both scenarios, the corporations lose.

Getting Comfortable and Then Phasing Out Research   Success breeds not only a growing market but also a narrowing of focus and a complacency about sustaining research to foster breakthroughs. As a company succeeds with its products, it becomes vested in growing its product lines. Incremental improvements for satisfying current customers and increasing market share begin to matter most. Except for companies that have endured and learned from this cycle before, by the time companies have developed high-technology products they have usually entered a routinization phase. They focus on the short term and on improving existing products rather than on the long term and on creating new technologies. R&D organizations with enhanced capabilities for listening to today's customers usually have diminished capabilities for breakthrough research. Routine research and breakthrough research require different, almost opposite, cultures of innovation.[21]

Hennessy[22]:

Disruptive technologies cause problems for existing companies because they disturb their whole product line, their margin assumptions, and what they are doing. One of the reasons that startups have been successful is that established companies are reluctant to embrace that much change.[23] The analogy that I give is that they are reluctant to shoot themselves in the foot, although that is a lot better than what will happen to them later.

One of the indicators that a company is about to go into a long downward slide is when they start thinking about preserving their old customer base more than about attracting future customers. I have seen this happen in lots of computer companies. It is the beginning of the end because it means that they are basically trying to build a wall around what they have as opposed to getting more people into their camp. That is a failure in a company's ability to look far enough ahead to see that there is a new world forming out there and that they will have to play in it and figure it out.

What happens when a company hits the top of its S-curve is analogous to what happens in a military situation when there is a surprise attack. Things seemed to be going along fine, then suddenly the market becomes saturated or a competing product seems to come out of nowhere. Stanley Feder describes this situation as follows: "Customer orientation is [a] mindset, and one that can cause large companies to ride existing technological horses off the cliff's edge. Instead it is the smaller rivals, free from the customer mindset, that often are better able to invest in next-generation innovations."[24]

Mindsets are easy to overlook. When a company gets good at making a particular kind of product for a particular kind of customer, some of its reactions become automatic. It develops special "lenses" for efficiently focusing on and seeing its customers. The fault lies in the dependence on the lenses. When the world changes, the company may not be paying attention in the right way. Research is one of the means by which a company can develop new options outside its current business.

When Research Gets Away   The second dilemma is faced by companies that sustain investment in research. A universal axiom is that breakthroughs contain major surprises. Even a breakthrough that occurs in the company's business area may interfere with or displace existing products. And a breakthrough may have its greatest applications in unexpected areas outside the company's main business. When that happens, the company has to manage an internal struggle over resource allocation[25]—a struggle between growing the main business and starting something new. There are many vested interests in the existing products, few vested interests in a breakthrough.

One of the most vocal critics of sustained research funding was Gordon Moore, long-time CEO of Intel. Moore strongly advocated that companies not invest in research. He believed that basic research did not directly benefit a company, and that most of the new ideas developed by research would not be usable by the company.[26]

To reap the benefits of an ill-fitting innovation, a company must either license the technology to another company or spin off a new company (perhaps with partners). Since these strategies are outside of its core business, a company has less experience and assumes greater risks. Typical hazards for the company that made the research investment include executing the new business poorly and having the individuals who made the breakthrough leave to start their own business or join another company. The companies that sponsor breakthrough research don't necessarily profit from it.

The two dilemmas detailed here make sustained funding of research for breakthroughs a high-stakes bet with big risks and big potential benefits. From the perspective of the larger society, funding breakthrough research is a good investment because *somebody* (and often many people) will benefit when products are developed.[27] However, from the perspective of an individual company with a narrow market and a specific product line, funding breakthrough research can be more akin to gambling. Even if the new technology pays off big, in the absence of effective strategies the benefits may go to other companies.

At the core of both dilemmas is the challenge of managing the costs and benefits of innovation and breakthrough research. On one path, a company may wind down research once it has launched a successful product; that then leaves the company with no capacity for generating more breakthroughs, which will be needed in the future. On another path, companies that sustain investment may not benefit from surprising breakthroughs.

Dealing with these dilemmas is the central challenge for the innovation ecology.

The Breakthrough Dilemmas for Universities

Bringing an insight to a point where it could be useful usually takes many years. Creating a technology from research done by students is problematic because students are short-timers. By the time they develop their skills, they are ready to graduate. Even when a university comes up with a usable technology, it has to find an effective way to transfer it to a product organization. (Universities do not have product organizations; their mission is education.)

Another dilemma for universities is that the knowledge needed to create breakthroughs—especially for radical innovations—often does not fit neatly within the boundaries of a single academic discipline. For example, advances in semiconductors have required intense collaboration not only in fabrication and material science but also in system design, optics, and imaging. Breakthroughs in biotechnology and medicine increasingly require not only multiple areas of biology but also engineering and computer science.

Universities are mainly organized around departments representing specific fields. New professors seeking tenure in a particular department take a big risk if following a problem to its root takes them too far from the interests of their home departments. This tends to limit cross-disciplinary research to professors who already have tenure.

For students pursuing doctoral degrees, the multi-disciplinary quality of breakthrough research creates special barriers. Institutional structures at universities are optimized for students pursuing a degree in a single field. Different fields have different methods, different knowledge, and different evaluation criteria. When a doctoral research project crosses or combines fields, the arrangements of thesis committees and funding support fall outside the standard modes and require more negotiations.[28] If the research is carried out by several students in a multi-disciplinary team, there is the added complexity of coordination and credit assignment. Finally, graduates whose work crosses and combines fields may find that this creates obstacles. Field-centric recruiting committees find that the work is outside their expertise or that the qualifications of the cross-disciplinary candidates fall outside the consensus-determined needs for the department.

The organization of universities into departments fits the historical pattern according to which most academic research is basic research. For this reason, most of the training at universities takes place within departments.

The Breakthrough Dilemmas for Federal Granting Agencies
In the United States, federal agencies also have breakthrough dilemmas in funding and managing research programs. These programs generally receive more proposals than they can fund and need to select projects by fair evaluation criteria. When agencies try to spread parts of a project across multiple contractors, this counters charges of favoritism but can impede progress in research because it drastically increases the overhead of collaboration in multi-disciplinary projects.

The evaluation criteria for basic research and those for applied research are quite different. Basic research programs are supposed to create new knowledge. Following academic practices, these are usually organized by discipline, and proposals for research are usually judged by people within a discipline. Because breakthrough research is often multi-disciplinary, it tends not to fit well into basic research funding programs.

Applied research programs are intended to solve known and important problems. Because of the sense of urgency, applied research projects are judged by whether they can make rapid and effective progress. This drives applied research in the direction of low-risk and incremental approaches. There is little room for experimentation or for theoretical work in applied research. Because breakthrough research typically requires the creation of new knowledge by novel approaches, it often runs counter to institutional expectations for applied research. The challenge for federal granting agencies is to find ways to fund research that is likely to lead to breakthroughs.

A Brief History of Corporate Research

It is said that one of the important inventions of the 20th century was the corporate research lab.[29] Corporate labs powered the explosive innovation that characterized the century. With globalization and other social changes, corporate research labs are now evolving with the rest of the innovation ecology.

The first era of industrial research organizations in the United States (1880–1906) coincided with the rise of big business. At the beginning of the 20th century, a few large companies, including General Electric, American Telephone and Telegraph, Dupont, and Standard Oil, opened laboratories to create technical and business advantages. The GE lab, founded by Charles Steinmetz, was the most famous. It became known as "The House of Magic." Its staff grew from eight in 1901 to 102 by 1906. About a third of the staff members had scientific training. After accounting, research was the first centralized function in business. The GE lab's main role was to keep the company informed about new discoveries in science and technology. By the early 1900s there had been much progress in basic science. Before the creation of corporate research institutions, most scientific research was done at universities. Little attention was given to commercial applications. The first corporate research labs were created to fill the gap between scientific insights and the creation of technologies for industry.

The second era of industrial research (1906–1926) was affected by World War I and by the rise of international competition. During the war, industrial research was stimulated by an infusion of government funding and a moratorium on patent controls. During this period, there was also much growth in federal funding to state universities. The universities focused on the application of science and engineering. Many of the companies, especially in the chemical and metallurgy industries, gained commercial advantages from the development of plastics, synthetics, and advanced metals during the war. During this period there was a national emphasis on quality and productivity. The National Bureau of Standards,[30] founded in 1901, coordinated research being done in companies with that being done in research institutions and that being done at universities. Companies built up excess production capacity during the war, and one role for the laboratories was to create new categories of products that could use this capacity. Many foreign-trained researchers were "imported" to staff industrial labs.

The third era of industrial research (1926–1950) was affected by the Great Depression and then by World War II. The improvements in productivity for industrial processes that had helped in World War I led to market saturation and loss of jobs during the Depression. There was a widespread perception that humanistic studies were needed to find ways to catch up with the advances in scientific discovery. The purpose of research shifted toward creating new businesses. The competitive business environment became more complicated. The new technologies created by research dominated the market, and competition was limited by the enforcement of patents. Antitrust legislation was introduced to counter the economic effects of the concentration of patents, sometimes by one company in an industry. Federal funding for research grew from $48 million in 1940 to $500 million in 1945. The scientific benefits of World War II were felt mainly in the industries that produced aircraft, electrical goods, and instruments. The role of science in ending the war was very clear in the public mind. There was broad public interest in funding science. The National Science Foundation, a major institution for government funding of research during peacetime, was founded in 1950.

In the fourth era of industrial research (1950–1975), the role technology had played in World War II increased the prestige and value of science in the public eye. Defense-related agencies such as the Office of Naval Research, the Atomic Energy Commission, the Defense Advanced Research Projects Agency, and the National Aeronautics and Space Administration allocated growing amounts of money to research. Companies set up new research facilities to broaden the scope of their activities, and secretive national laboratories were founded to focus on defense technologies. There was a national emphasis on research self-sufficiency and manufacturing optimization. Basic research, perceived as a driver of economic growth, attained a high status. The "brain drain" of scientists from Europe continued.

In the 1960s, there was a fresh infusion of government funding into research after the Soviet Union launched the first man-made satellite. During this period, several major scientific discoveries were made in corporate research laboratories. For example, Arno Penzias and others at Bell Labs discovered dark matter, and superconductivity was discovered at IBM. Corporate R&D attracted the brightest people and provided advanced equipment to those working on cutting-edge problems. New business competitors faced substantial barriers to entry, and companies with dominant market shares worked to stage the introduction of new technology in ways that would create obstacles for their competitors.

The fifth era of industrial research (1975–1990) was an era of rising international competition. Increasingly, companies began moving manufacturing overseas. Corporate research became decentralized and refocused on applied research and getting products to market faster. In this context, basic research organizations were increasingly seen as remote and irrelevant to corporations. Research managers had more difficulty getting innovations based on decades of basic research to market. Companies focused on growing their financial returns in existing markets, and industrial research activity declined in the early 1990s.[31]

The sixth era was shaped by the continued growth of international business competition and by the Internet bubble. Large companies adopted a more global perspective, and in the 1990s this led to increasing consolidation through mergers and acquisitions. This coincided with the establishment by international companies of multiple internationally located research institutions. With the increase in international competition, the know-how for commercial applications became more diffuse as more companies around the world began to open research laboratories. In addition, small companies increasingly entered the innovation ecology. Although they lacked the capacity to fund long-term basic research, they had the advantage of being able to evolve their business models rapidly. Cisco out-innovated Lucent not by doing its own research but by partnering with startup companies and acquiring technology from them.

In the heated economy of the 1990s, many companies emphasized quick results over innovation. Corporate management grew and became increasingly cumbersome. In an effort to counter this, there was a growing espousal of "intrepreneuring," the idea that organizations within a company should take the lead in developing new products and markets for them. Success in these endeavors was hampered by the absence of an understanding of the resources or the business skills required. The growth of the Internet bubble in the mid 1990s coincided with a growing understanding of how big companies can become structurally incapable of leading or embracing innovation and technological change. This fueled interest in entrepreneurship outside of big companies. As the economy collapsed, many people came to understand that small companies lacked the staying power and resources to develop markets.

In summary, the institutions for industrial research co-evolved with the fortunes of corporations through the 20th century. New technologies from corporate research laboratories created growth opportunities for the first big technology companies. This organized approach to industrialized science worked very effectively. By the middle of the century, the patent and know-how advantages of companies that had research laboratories led to concerns that science and patented technology were creating monopolies. Balances were sought to ensure the public good.

In the 1990s, globalization, short-term expectations, and maturing technologies began to change the game. Investors became less patient as companies expected quicker returns. Companies began relying more on outside sources for new technologies. Increasingly international competition coupled with maturing technologies led to lower profit margins and a focus on short-term profits. Companies with less cash found themselves in need of breakthrough innovations. As the 21st century gets underway, new strategies for creating technologies and breakthroughs are needed.

Strategies for a New Century

As the 20th century closed, the innovation ecology was in transition. Many of the big companies in the information sector were scrambling. There was a lively debate about the end of the personal computer era. However, the underlying issues were broader than the question of whether "personal computers improve productivity." Broad structural changes were taking place in all businesses based on semiconductors, modular manufacturing, and computation, including the computer business, the consumer electronics business, the video game business, and the business of telephone systems. Companies in all these business were going through the same kinds of transitions.

How will innovation fare in the new century? How will it fare across industry sectors, not just in the information sector? How will it be affected by increased globalization and the restructuring of industries? Will the effects of the breakthrough dilemmas leave us unprepared for the challenges and surprises that lie ahead?

Some properties of innovation and the innovation ecology will continue as they are now. Universities will continue to carry out basic research, funded by government and private grants. Government agencies will try to balance competing social needs, and will be challenged to provide stable funding for long-term projects. Venture capitalists will still fund startups and provide seed funding.[32] There will still be a gap between what basic research creates and what venture capitalists or product organizations need. We will still need breakthroughs.

Radical Research

To understand how innovation is changing, let us focus on the breakthrough zone and on what it can tell us about invention and innovation. The dominant pattern of innovation suggested by the S-curve in figure 1.2 begins with basic research to create new knowledge, which is followed by applied research to create new technologies and then by product development. A deeper look at examples of breakthrough research, however, shows that the middle part of this curve—the part represented by the breakthrough zone—is often characterized by an approach to research that is different from either applied or basic research.

If basic research had a slogan, it might be "Follow your curiosity wherever it leads you." Basic research is about creating new knowledge. It is guided largely by a sense of where nature's secrets will yield to scientific investigation.

If applied research had a slogan, it might be "Focus on the important problem. Don't get distracted by your curiosity." Applied research is about using what is known to solve problems. It is guided by a sponsor or client's sense of what problems are important. In its pure form, applied research does not take time out to investigate interesting "diversions."

Some of the most productive research done in the last century was neither basic research nor applied research. This alternative form of research—*radical research* or "following a problem to its root[33]"—can be a fast track to breakthroughs. It starts out like applied research when a researcher or a small team begins working on an important problem. An essential part of the approach is that the problem be important. The importance does more than ensure that the effort of finding a solution won't be wasted; it also motivates and guides the researchers.

If radical research had a slogan, it would be "Focus on an important problem, and follow that problem to its root." Radical research is guided by the problem and by the obstacles that arise.

In applied research, if there is an obstacle, the group tries to get around it. If stuck at an obstacle, the researchers look for a quick fix or give up. Applied research does not "waste time" trying to understand why something works or doesn't. In radical research, however, obstacles focus the research. Typically, a multi-disciplinary team is deployed to find perspectives on the obstacle. As the exploration deepens, more disciplines may be brought in as needed. This part of the research is like basic research, in that the energy is focused to create new knowledge.

Radical research can be strikingly efficient. Single-discipline projects judge which problems that they can tackle through the perspectives of one field. For example, someone specializing in optics would not plan a project to cure cancer. In contrast, radical research can tackle problems that don't fit neatly into a single discipline. Discoveries and insights along this journey often happen at the edges of disciplines or in what John Seely Brown calls the white space between disciplines. Following a difficult problem to its root often yields solutions that had eluded people before—that is, breakthroughs.

Open Invention and Open Innovation

Two questions are at the core of radical research: "What is needed?" and "What is possible?" Much of this book is about the interplay or dance between these questions.

To put it succinctly, radical research creates breakthroughs because of its efficiency in the collision and interplay between the two questions.

Broadly construed, product development is mainly concerned with "What is needed?" Its attention is driven to this question by its focus on addressing particular needs and solving particular problems. It pays little attention to "What is possible?" In contrast, basic research is mainly concerned with "What is possible?" Its attention is driven to this question by virtue of following curiosity and the quest for new knowledge.

There is a gap between the two questions. The results created by basic research ("What is possible?") are not ready for use in product development ("What is needed?"). Similarly, the insights gained by people trafficking in customer needs and emerging markets are not often considered by people in research. Much of the challenge in managing innovation and creating breakthroughs is about that gap. Radical research bridges the gap. It has all the sparking and zapping efficiency of a short circuit that puts both of the questions on the table at the same time.

Federal granting agencies and universities focus on the "What is possible?" side of things as they try to discover new knowledge. Growing companies and venture capitalists focus on the "What is needed?" side of things as they try to develop or discover markets. Corporate research is in the middle. That's why it is most often the home for radical research. However, when the economy is in structural turmoil it is more difficult to sustain a level of funding for bridging the breakthrough zone. Without stable funding, the breakthrough zone turns into "the innovation gap." Science and society's needs can not connect very well.

The challenge for the innovation ecology is to find stable ways to keep the breakthrough zone vibrant and productive. This productivity depends on invention (creating prototypes of new things) and on innovation (taking prototypes all the way to product, and developing new markets).

Looking at several case studies, Henry Chesbrough[34] suggests that efficient strategies for innovation essentially create more efficient markets for technology.[35] Increasingly, companies cannot create all the technologies that they will need. They need strategies for acquiring some technologies from other companies. They also need strategies for marketing their own technologies to other companies. Chesbrough characterizes these kinds of strategies under the rubric of "open innovation." He advocates open business practices for promoting and acquiring innovations.

Chesbrough's insights seem exactly right, but they address only half of the equation. They address strategies for efficient *innovation*, but not strategies for efficient *invention*. In a business climate where technology companies are building products on the same technologies and where they are outsourcing manufacturing to the same outside vendors, open innovation is not likely to save the day. Companies are already practicing open innovation. Their products are turning into commodities and their profit margins are shrinking. What open innovation misses is creating a path to renewal. Breakthroughs are needed. *No one ever created a breakthrough from open innovation.*

The dialogue between "What is possible?" and "What is needed?" requires efficient conversations between scientists and inventors on one side and marketers on the other. In short, creating breakthroughs requires not only "open innovation" but also "open invention."

2

The Dance of the Two Questions

Two roads diverged in a wood, and I—
I took the one less traveled by,
And that has made all the difference
—Robert Frost, from "The Road Not Taken"

Traditionally, the word 'innovative' has been used in sentences such as "That's a very innovative solution to a tough design problem" to mean *novel and creative*. It connotes "thinking outside the box." However, in recent years 'innovative' has been increasingly used to make an important distinction from "mere invention." Invention is the first part where something new is created. Innovation is taking an invention *all the way* to a product.

Every successful innovation requires addressing two questions: "What is possible?" and "What is needed?" These questions[1] are at the heart of invention and innovation.

"What is possible?" concerns research, discovery, and invention. "What is needed?" concerns business and social needs. Because these questions pull in different directions, deciding what to do in research can be confusing. Nonetheless, for an innovation to come into the world both questions must be asked and answered. Together these questions address the means and the ends of innovation.

Stories

What Is Possible?
In June 2002 the editors of *IEEE Spectrum* published an interview with Herbert Kroemer after he received the IEEE Medal of Honor.[2] Earlier,

Kroemer won a Nobel Prize with Zhores Alferov for developing semi-conductor heterostructures[3] used in high-speed electronics and in opto-electronics. Kroemer reflected on his long research career and the nature of his work. Although ultimately his research led to important products that millions of people use and depend on, Kroemer always focused more on science than on innovation. In the *Spectrum* interview, he said "Certainly when I thought of the heterostructure laser, I did not intend to invent compact disc players. I could not have anticipated the tremendous impact of fiber-optic communications. The person who comes up with the applications thinks differently than the scientist who lays the foundation."[4]

As the editors of *Spectrum* noted, Kroemer's career does not reflect a person trying to invent. Kroemer did not set out to solve a particular problem, meet development milestones, or follow formal development processes. Rather, his work was driven by curiosity and a spirit of inquiry. As they put it, Kroemer "reminds us that transistors, lasers, and fiber optics did not come out of Six Sigma or ISO 9000 certification quality control programs."[5]

Kroemer believes that new technologies are mainly for solving new problems. It is futile to try to predict the specific use of new discoveries in terms of known problems. He encapsulates this notion in what he calls Kroemer's lemma: "The principal applications of any sufficiently new and innovative technology always have been and will continue to be applications created by that new technology."

In simpler language, Kroemer's lemma says that new technologies solve new problems. This lemma rings true for two reasons. One reason is that the understanding of what is possible opens doors to new kinds of applications. Another reason is that the problems of the future can be difficult to discern in the present. You can't easily ask a future customer what they will want. Relying on what a customer wants *today* is too simplistic for breakthrough inventions and discoveries. The problems that *will be* most important in the future are difficult to recognize in the present. When inventions may take years to develop, an inventor must begin working on the solution long before any real customer would show up.

Kroemer's lemma rings true, but it is only part of the story. Among perspectives on innovation, it represents an extreme point that gives primacy to the question "What is possible?"

What Is Needed?

Ted Selker has invented many things, especially devices for interacting with computer systems. Selker is known as a hyperkinetic wild man. His eyes sparkle, he talks rapidly, and his words can't keep up with his brain. He has more ideas in a day than many inventors have in a decade. When we asked Selker how he picked problems to work on, his answer was startling[6]: "Actually, I don't consider a problem interesting or worth working on unless I have a customer who wants it solved. I want to start with solving a real customer's real problem." This problem-solving perspective is essentially the opposite from what was voiced by Kroemer. Kroemer asks "What is possible?" and Selker asks "What is needed?" If Ted Selker had a lemma, it might be "Don't start with a solution looking for a problem. Start with a problem and then find a solution for it."

Selker's approach rings true for several reasons. For many researchers, knowing that a solution would have great value is motivating. For example, some medical researchers are deeply motivated in their search for cures for diseases that cause much suffering. Starting with a problem sidesteps the difficulty and expense of working for long periods of time to solve a problem that is not worth solving. When a potential solution for a hard problem already fits within the general business of a sponsoring company, there are further advantages. Researchers have powerful arguments for attracting funding, and if they find a solution they will not have to search for a company to develop a product based on the invention.

Selker's lemma rings true, but again it is only part of the story. Among perspectives on innovation, it represents an extreme point that gives primacy to the question "What is needed?" The real action in innovation requires navigating between these rhetorical extremes.

The Dance of the Two Questions

Not all inventors begin with *one* of the two questions "What is possible?" and "What is needed?" There is a dance between the two questions as an invention comes into focus. The following stories illustrate different ways of grappling with the two questions. Cross-disciplinary teams sometimes start with a problem ("What is needed?") and then use specialists in different disciplines to understand how to crack the problem ("What is possible?"). The dance between the questions can be intricate and involved.

Cross-Disciplinary Teams   The EARS project[7] at Xerox PARC built the first laser printer. Chuck Thacker was on the team. From his perspective, it was a small team of highly competent experts—each willing to drive the research in his field toward achieving the common goal of creating the laser printer[8]:

EARS is one of the best examples of the cross-disciplinary projects that actually worked and made Xerox a lot of money. It was the first laser printer. EARS is an acronym for Ethernet, Alto, RCG, and SLOT.[9] People from both the computer and the systems lab built EARS over a very short period and it was really quite difficult. The result was a tour de force. The company did not quite get it for several years, but they certainly made a lot of money on it.

This laser printer could print 60 copies a minute at 600 dots per inch. Ron Rider was the primary leader of that project. That was a perfect example of the kind of cross-disciplinary project that worked wonderfully. We had Gary Starkweather beavering away on the lasers and the optics [figure 2.1]. Ron Rider and Butler Lampson worked together on the character generator. I was working on the Alto. The computer networking was mainly done by Bob Metcalf and David Boggs. Software was written by a wide variety of people. It was a ten-person project from all over the place, with a lot of different specialties and it worked like a charm.

No one had ever built anything like this before. We had built some raster printers earlier based on a fax technology that Xerox invented in the mid sixties. EARS was harder because putting down a 600 dpi[10] page every second required enormous bandwidth for the time. It would be the equivalent of an IMAX theater today. That kind of bandwidth, the custom logic that ran the printer was the equivalent of a graphic super computer and there were a lot of problems. The algorithms were hard. How do you get a gazillion fonts? How do you get the fonts necessary for the next page that is about to be imaged in memory now so that they will be there when they are needed? There were a lot of serious technical challenges in the design of that system. It worked just fine. Communication was crucial because Gary would want to make a change to the way that the scanner worked and that would change the clock rate of everything, but it got done.

There have been very few projects like that in my career that I have been as satisfied with. This was absolutely a small crack team. Don't leave home without one! (laughter) We had all known each other for a long period of time. People at PARC and DEC[11] are uniformly pretty smart and uniformly pretty trustworthy. You don't have any trouble taking on a project that involves a short jaunt down a dark alley with those kinds of people.

The laser printer was created by a multi-disciplinary group, working together to find a solution to an important problem. It was a major

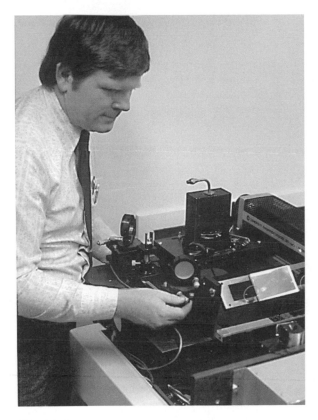

Figure 2.1
Gary Starkweather working on an early laser printer. Courtesy of Palo Alto Research Center.

breakthrough. The twists and turns of their work reflected the journey of a group following a problem to its root.

Following a Problem to Its Root   John Seely Brown sees the choice of a difficult but important problem as essential to the success of innovation, including the laser printer and many other breakthroughs.[12] Brown defines "radical research" in terms of the method of following a hard problem to its root:

When we get in the spirit of following a problem to the root, that pursuit of listening to the problem brings multiple disciplines and multiple crafts together. The *problem* pulls people together.

If you call a meeting and say "I want to have a sociologist, a psychologist, and a physicist come to this meeting." I'll guarantee you nothing will happen. But if there's a problem that you have identified which requires triangulating in the pursuit of a solution—researchers get sucked into it. The problem pulls us out of our individual disciplines and causes us to leave them behind and cross field boundaries. That's where the productive clashes of craft really start to happen.

Unlike basic research, radical research is informed by a problem to solve. Radical research is energized if the problem space you are going after is important. It's not just something that your academic community tells you is important. It's a problem that is situated in the world and needs to be cracked. So it's a real problem with real significance.

If you follow it to the root, it is inevitable that not only will you find something that is of tremendous value in the space where you started out, but also you will find all kinds of other value. And so you will need ways to monetize or capitalize on all the different ways that a fundamental breakthrough will lead to.

One example is solid-state multi-beam lasers. That research led to an empire of uses. We started the laser work to drive printing in Docutech.[13] The solid-state lasers enabled printing at twice the speed and at a tenth of the cost of gas lasers. But solid-state lasers have many applications beyond laser printing and that's why we created SDL [a laser company].[14] We wanted to find a cheap way to manufacture the lasers and to amortize the cost of manufacturing them. When SDL was finally sold [in the late 1990s], it had ten times the market capitalization of Xerox.

Suppose that you are doing applied research. If you hit an unbelievably big barrier, you are depressed. You work as hard as you can to get around it. If you are engaged in radical research, the bigger the problem you hit, often the more interested you are because maybe you've stumbled into something fundamental about nature as you go to the root of that barrier to understand it. So applied research circumvents obstacles and radical research goes to the root. Radical research leads simultaneously to tremendous new knowledge *and* something that has great utility, often utility far beyond what you first had in mind.

The pattern of radical research was not invented by John Seely Brown or even at PARC. It is a style of research whose practice goes back to the earliest times of science. It is followed in many research organizations and universities. Variations of it appear in many engineering departments.

Radical research—the approach of following a problem to the root—addresses inherent difficulties of both basic research and applied research. A known limitation of applied research is that it is usually locked into incremental improvements and does not tend to create breakthroughs. Basic research can create breakthroughs, but often in areas that have little intrinsic value, at least in time frames of a few years. Basic research requires great patience to yield results. It is also hard to accept the uncertainty about what the applications of the results of basic research will be. In contrast, radical research can yield the best of both worlds. It starts with a known problem—so if a solution is found, it is already known that it could have great value. Furthermore, by potentially creating new knowledge, it avoids the shortsightedness of incremental improvements and can potentially lead to breakthroughs.

Finding the Problem   Finding a customer's real problem requires listening, and listening requires time and networking. Henry Baird, one of the inventors of CUDA[15] (a system for scanning documents, especially legacy documents, and making them readable on small hand-held devices) had this to say[16]:

The idea of CUDA is that we take images of printed text and represent them on small format displays such as PDAs [personal digital assistants] or small windows in such a way that the text images re-flow in much the same way that HTML[17] re-flows in web browsers. The text re-justifies itself and fills the available space, preserves fonts, and is highly intelligible. The idea is to make beautiful printed text available to people on multiple types of displays.

CUDA came about as a consequence of collaborations. I make a point of networking and going up to the University of California at Berkeley on spec without any particular agenda, listening very carefully and getting to know everyone. I touch base with everyone, following people's interests and gathering raw material very deliberately without a focused goal.

CUDA arose from a combination of four or five originally very distinct research ideas. The pieces were as follows. The basic expertise in document image processing was Dan Bloomberg's, and basic expertise in software related to PDA devices was Bill Janssen's, the PARC book scanner effort was under Bob Street and Steve Ready, and the Digital Library Project was at the University of California at Berkeley. So I was in touch with all of these efforts.

Dan Bloomberg worked for me. Bill Janssen was a guy in another group that I occasionally spoke to. The PARC book scanner people were people that I had introduced myself to because we had been doing some software for them. I also was a member of the Digital Library Project at Berkeley.

The following unplanned sequence of events occurred. The book scanner people were looking for an early, large-scale trial of their scanner. The University of California at Berkeley people were looking for a way to scan in certain rare and fragile and yet nevertheless heavily used scholarly reference works. They wanted to put them up on their digital library site and this would involve many thousands of pages. So it was a scale of effort that they could not easily accommodate without assistance.

The people from Berkeley selected some books to scan, including a book called the Jepson *Flora of California*, which is the earliest and still most authoritative description of botanical species in California. Berkeley actually holds the copyright on that book. They are seen as the source of that book. It was not available in digital form. It's a very long book of 2,500 pages. And so it was very natural for them to publish it on their web site.

So the Berkeley people sent staff down here to PARC to run the book scanner. Out of that came statistics about how well the book scanner works. The collection of Jepson images was exciting to everyone involved and to me in particular because it gave us the raw material for cutting-edge document image analysis work. That was the first stage.

Then independently of that Dan Bloomberg's father-in-law was having trouble with his vision and reading was difficult. Dan came in one day and was all excited and said "I'm going to build a system called BigType. I'm going to scan an image of a book and re-typeset it with very large print without doing OCR (optical character recognition) on it. I'll blow up the text so that he can read it. I'll take a book that's 100 pages long and reprint it so that it's 500 to 1,000 pages long with very big type and my father-in-law will be able to read it." Dan was very excited about this.

For the moment these ideas did not connect to one another at all. Then independently of that I learned that Bill Janssen was interested in PDAs. He'd come round showing us his advanced PDAs, talking to us about them. That just seemed to me like an interesting curiosity.

The key that brought the ideas together was a remark by Tony Morosco, a professionally trained botanist working on the Digital Library Project at Berkeley. He said "I'm very happy to have these 2,500 pages of the Jepson *Flora* on hand. Now the botanists are telling me that they would like to have them *in the field* as they go around trying to identify plants."

Tony was also connected to the Jepson Herbarium, which is the place where the Jepson *Flora* was originally written. So he had been talking to botanists, who are the potential end users of the system, and they said "What I would like to do is to access the Jepson *Flora* over my PDA as I'm staring at some unknown bush."

And all of a sudden the pieces clicked and this was for me the moment of creation. I realized that we had the Jepson *Flora* in image form, and Dan Bloomberg's BigType idea could be applied to it in principle, to make it possible to display in very large type in a book form. But exactly the same technology would work to shrink the representation down to fit in a PDA screen.

The ideas came together when I was driving back from Berkeley. I got so excited that I almost drove off the road!

Like the story of the laser printer, the story of CUDA is about a multi-disciplinary team. The CUDA story, however, is not one of a team following a given problem to its root. Before a team can chase after a qualified problem, the problem needed to be identified. The first step in Baird's dance was to find a problem representing "what is needed."

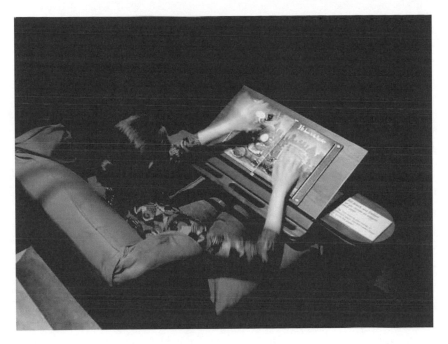

Figure 2.2
Maribeth Back using a ListenReader. Courtesy of Palo Alto Research Center.

Baird's pattern is to marinate in a set of problems and technologies until a particular opportunity emerges.

Improvising Toward Clarity   In the next example, the researchers did not follow a problem to its root, nor did they follow a technology to a problem. Rather, they danced quickly between "What is possible?" and "What is needed?" to define a project. This pattern is characteristic of approaches grounded in design.

Maribeth Back, an electrical engineer with strong interests in theatre and music, worked at PARC for several years. Her invention, the ListenReader, was a popular part of the "Experiments in the Future of Reading" exhibit at the San Jose Tech Museum in 2001. The Listen-Reader looks like a regular book (figure 2.2) However, the book is wired—it contains hidden sensors and connections to a computing system. As readers of the ListenReader move their hands or touch the pages, it becomes interactive, triggering sounds and other effects that enhance the story.[18] Back:

I know that creativity works in different ways for different people. For me, I have to work on four or five projects at once and somehow the ideas out of those projects interweave with each other and enrich each other. And the project may turn out to be completely different from what I expected in the beginning.

My current favorite invention is the ListenReader. It brings together all of my loves. I love reading. I love doing sound design. I love doing theatre. And it brings all those things together. It creates a personal theatrical experience for the person sitting in the chair. So the person reading the book is an artistic participant, engaging in a personal performance using the materials provided by the author and the sound designer. It's not like interactive narrative, where the reader must make conscious choices as much as it is like the experience of a person in a live theater audience. The audience's presence and vitality are integral to the experience, but do not necessarily direct the story flow. So for me, it's sort of like encapsulating all the work I've done in the past in radio, in theatre, this life-long love of reading that I've had. Bringing it all together and wrapping it up into one ideal capsule—a little encapsulated theatre, a personal theatrical device.

The project actually came from some work I'd done on the Brain Opera, which was an electronic interactive opera done at the MIT Media Lab. I was the sound designer and a performer for it. They had worked with capacitive field sensors. They are magic sensors. You wave your hand in the air and something happens. It's like Mickey Mouse in *Fantasia*.

I wanted to take these sensors and make them into reading devices. We'd used them for music in the Brain Opera, so I understood their parameters very well. I knew how to make them work. I knew how to program them. I wanted to combine them with something besides music.

This work was done in early 1997. I thought "Why don't I build the sensors into a book and have the book be the controller?" I really had the Mickey Mouse *Fantasia* image in my head when I was thinking about that: the book as the magic controller. I thought it would control some kind of dynamic system such as lighting. We were actually talking at some point about having it control air. So it would blow warm and cool air at you and you'd have a real physical experience from the turning of pages of the book. I did not know that it would be controlling sounds.

In summary, Back does not start by understanding a problem to be solved or the technology that could be used. She is inspired by an abstract version of each. She co-evolves the design through a process of rapid prototyping until she has a solution that solves a problem that becomes identified along the way.

Reflections

Patterns of Attention

Inventors do not all work the same way. Figure 2.3 shows how different inventors discussed later in this book shift their attention between

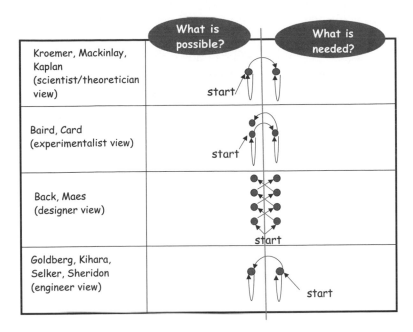

Figure 2.3
Patterns of attention to the questions "What is possible?" and "What is needed?"
Arcs and lines with arrowheads represent transitions from one focus of attention to another and transitions to another aspect of the same focus of attention.

the two questions. The bottom row of figure 2.3 shows that Nick Sheridon, David Goldberg, Nobutoshi Kihara, and Ted Selker use a problem-solving perspective and start with the question "What is needed?" This approach is the dominant method for people with an orientation of engineering.

The diagrams in the top row in figure 2.3 show how some inventors, like Herb Kroemer, start with the question "What is possible?" Jock Mackinlay is known in his work for developing a technology and then performing a systematic search for applications of it. This approach can be efficient when most of the work is in developing a broad technology with many possible uses. This approach is the dominant method for people with an orientation of science, especially theoreticians.

The third row in figure 2.3 includes the pattern of attention in Maribeth Back's story of the development of the ListenReader. We call the back-and-forth movement "the dance of the two questions."

Breakthroughs often arise from a multi-disciplinary team, as in Thacker's laser printer story. In following a problem to its root, the team can change directions as it deepens its understanding of what is needed. Different members of the team may have individual patterns in different quadrants, but the team as a whole also has a global pattern. When teams divide the work into subprojects, each subproject brings specialized knowledge for discerning what is possible. Collectively, they address both questions. In summary, the rhythm and step of the dance may vary, but innovation must ultimately reconcile its answers to both "What is possible?" and "What is needed?"

Science Policy and the Two Questions

Simple as they seem, the two questions are profound and the interactions between them are subtle. Debate about these two questions has a prominent place in the social history of science and technology. Often the question "What is possible?" is ascribed to the domain of "basic research" while the question "What is needed?" is ascribed to the domain of "applied research."

Near the end of World War II, a perspective on these two questions took hold in the United States and shaped science policy for at least three decades. Vannevar Bush, director of the wartime Office of Scientific Research and Development, was asked by President Franklin D. Roosevelt to plan ahead for the role of science in the period of peace that was expected to be ahead. Bush wrote a report titled *Science, the Endless Frontier*. This report laid out a perspective for investing in basic research. It framed how basic research could lead to social benefits. This view became the foundation of U.S. science policy for several decades.

In his book *Pasteur's Quadrant: Basic Science and Technological Innovation*, Donald Stokes brilliantly analyzed the influence of Bush's work. The heart of the analysis arises from two aphorisms in Bush's report. The first was that "basic research is performed without thought of practical ends"—essentially the "What is possible?" theme. Bush went on to characterize basic research as contributing to "general knowledge and an understanding of nature and its laws." Basic research is the pacemaker of technological progress. Stokes summarized Bush's view as follows:

If basic research is appropriately insulated from short-circuiting by premature considerations of use, it will prove to be a remote but powerful dynamo of technological progress as applied research and development convert the discoveries of basic science into technological innovations to meet the full range of society's economic, defense, health, and other needs. (p. 3)

Bush went on to assert that "a nation which depends upon others for its new basic scientific knowledge will be slow in its industrial progress and weak in its competitive position in world trade."

The end of World War II was a significant moment for framing science policy. Science and particularly physics were in the public eye because of their dramatic role in creating the atomic bomb and bringing about the close of the war. One can imagine that the experiences and emotions of scientists and others during this period in history were very complex. Scientists were conflicted in their attitudes between embracing and distancing themselves from their roles in the war effort. Some of them wanted to get back to a quieter version of research without the secrecy or even guilt of wartime applications. Stokes quotes J. Robert Oppenheimer, director of Los Alamos Scientific laboratory:

. . . the things we learned [during the war] are not very important. The real things were learned in 1890 and 1905 and 1920, in every year leading up to the war, and we took this tree with a lot of ripe fruit on it and shook it hard and out came radar and atomic bombs. . . . The whole spirit was one of frantic and rather ruthless exploitation of the known; it was not that of the sober, modest attempt to penetrate the unknown.

In the context of this period, Bush's report argued strongly for a separation of basic and applied research. This view of basic research—its nature, role, and necessary remoteness from application—had a big influence in framing U.S. science policy. It helped sustain an investment in research whose goal is to expand human knowledge.

As Stokes has noticed, the influence of Bush's report led to a one-dimensional view of how innovation happens. In the one-dimensional view portrayed in figure 2.4, basic research feeds results to applied research, which then feeds results to advanced development, which then

Figure 2.4
A linear or one-dimensional model of technological innovation.

feeds results to product development and operations. Everything flows in a line from basic research to product.

The Relation between Science and Technology

While this view of technological innovation is dogma in some circles, it is not without its detractors. In his book *Managing the Flow of Technology* Thomas Allen challenges the belief in this model directly:

Despite the long-held belief in a continuous progression from basic research through applied research to development, empirical investigation has found little support for such a situation. It is becoming generally accepted that technology builds upon itself and advances quite independently of any link with the scientific frontier, and often without any necessity for an understanding of the basic science which underlies it. (p. 48)

Allen goes on to cite the historian of science and technology Derek de Solla Price on this point:

Physical science and industrialism may be conceived as a pair of dancers, both of whom know their steps and have an ear for the rhythm of the music. If the partner who has been leading chooses to change parts and to follow instead, there is perhaps no reason to expect that he will dance less correctly than before.

Much of Price's study of science and technology has focused on the roles of technology and craftsmen in creating the laboratory instrumentation that enables science to forge ahead. Most observers of the overall scene doubt that the astonishing breakthroughs that fueled Vannevar Bush's conclusions near the end of World War II could have come about from a process of incremental engineering.

At the core of this debate are two fundamentally different ways of learning. Theoretical science looks inward for patterns and appreciates the value of simplicity. In the face of compelling evidence, theoreticians sometimes have more confidence in their models than in the vagaries of experimental data. In his book *Searching for Solutions*, Horace Judson quotes the physicist Murray Gell-Mann on this point:

You know, frequently a theorist will even *throw out* a lot of the data on the grounds that if they don't fit an elegant scheme; they're wrong. That's happened to me many times. [For] the theory of the weak interaction: there were *nine* experiments that contradicted it—all wrong. Every one. When you have something simple that agrees with all the rest of physics and really seems to explain what's going on, a few experimental data against it are no objection whatever. Almost certain to be wrong.

Technologists—whether we call them applied researchers, engineers, designers, or tinkerers—gain confidence in their ideas in a completely different way than theoreticians. They look outward rather than inward. The cognitive scientist and former director of PARC John Seely Brown put it this way: "There is a huge difference between theoreticians and inventors. Inventors are often tinkerers, and tend to get more visible feedback, more tangible support from the surrounding context."

The messages of Price and others tell us to celebrate the value of both sources of insight and their cross-fertilization.

Pasteur's Quadrant

Some of the best innovation has taken place when the two questions have *not* been isolated. As Stokes illustrated from the scientific lives of Pasteur and others, "the belief that the goals of understanding and use are inherently in conflict, and that the categories of basic and applied research are necessarily separate, is itself in tension with the actual experience of science" (p. 12).

As Stokes remarks, even as Pasteur laid out new branches of science, every study that he did was applied. Several Nobel laureates have expressed similar views about the link between understanding and use. Arthur Lewis made his most important contribution to development economics by probing the deepest intellectual puzzles in economics. Irving Langmuir felt that understanding the principles of the physical world and making improvements to technology were parts of the same venture.

The one-dimensional view leads to confusion at the institutional level. Research viewed as basic by a scientist may be viewed as applied by a sponsor. Research may even be perceived differently according to *where* it is done. According to Stokes, "certain types of research on semiconducting materials, carried out in a university laboratory, might be regarded as fairly pure, while in Bell Laboratories they would be regarded as applied simply because potential customers for the research results existed in the immediate environment." (p. 63)

Crucially, Stokes argued that the one-dimensional view fails to recognize that new knowledge arises not only from the pursuit of knowledge *without thought of use* but also from the pursuit of knowledge *with a clear view of its possible use*. Stokes proposed the two-dimensional model illustrated here in figure 2.5. In this model, the two axes

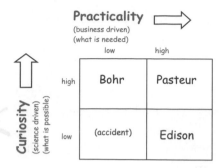

Figure 2.5
Donald Stokes's two-dimensional model of research and innovation.

correspond to the two questions. The vertical axis corresponds to "What is possible?" It reflects the degree to which the researchers or inventors—especially theoreticians—are driven by curiosity or the desire to increase knowledge and are guided by such considerations as elegance, simplicity, or beauty. The horizontal axis corresponds to "What is needed?" It reflects the degree to which researchers or inventors are driven by practicality and guided by economic or social values.

Following Stokes, figure 2.5 has four quadrants. The upper left quadrant represents research in which consideration of curiosity is high and consideration of need is low. It is named *Bohr's quadrant* after the eminent physicist Niels Bohr. This quadrant characterizes the pure scientist's or even pure mathematician's approach for pursuing knowledge without consideration of application.

Taken to an extreme, this perspective can be antithetical to innovation. C. P. Snow put it as follows in *The Two Cultures: And A Second Look: An Expanded Version of the Two Cultures and the Scientific Revolution*: "We prided ourselves that the science that we were doing could not, in any conceivable circumstances, have any practical use. The more firmly one could make the claim, the more superior one felt." (p. 32)

Among the invention examples in this chapter, this quadrant is closest to the perspective of Herbert Kroemer. In contrast to Snow's characterization of his colleagues at Cambridge, Kroemer believes that science *should* have applications. What he argues is that the two questions are best addressed by different kinds of people.

The quadrant at the lower right is named *Edison's quadrant* after Thomas Edison, inventor of the light bulb, the phonograph, and many other things. Although he published at least one important scientific paper on the "Edison effect" (a paper describing the basis of operation of vacuum tubes), Edison focused more on solving problems than on creating new knowledge.

The unnamed lower left quadrant corresponds to accidental discoveries, where neither question is in focus.

The upper right quadrant is named after Louis Pasteur. It is for research that finds scientific *inspiration* in problems that address needs. Stokes characterizes this quadrant as follows: "It deserves to be known as Pasteur's quadrant in view of how clearly Pasteur's drive toward understanding and use illustrates this combination of goals. Wholly outside the conceptual framework of the Bush report, this category includes the major work of John Maynard Keynes, the fundamental research of the Manhattan Project, and Irving Langmuir's surface physics. It plainly also includes the 'strategic research' that has waited for such a framework to provide it with a conceptual home." (p. 74)

An interesting study reflecting this framework, cited by Stokes, is a report by Comroe and Dripps on the developments in physical and biological science that led to advances in diagnosing, preventing, and curing cardiovascular or pulmonary disease. With the help of 140 consultants, Comroe and Dripps identified the knowledge essential to each advance and analyzed 500 key articles. Their analysis probed in part whether the research was clinically focused (addressing "What is needed?") or whether it was carried out with no consideration of clinical problems ("What is possible?"). They put 37 percent of the articles in Bohr's quadrant, 25 percent in Pasteur's quadrant, and 21 percent in Edison's quadrant.

The Two Questions in Daily Research Life

From his position as professor of politics and public affairs at Princeton University, Stokes had an interest in public policy. He studied how science and innovation were managed to benefit the social good. His thesis was that the one-dimensional view is suboptimal for guiding research policy.

Figure 2.6 relates Stokes's two-dimensional model to our observed patterns of attention, showing that the basic patterns of attention correspond directly to the three named quadrants in Stokes's model.

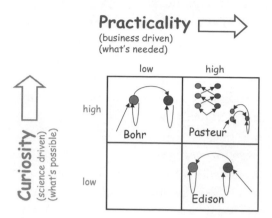

Figure 2.6
Patterns of attention in science and invention.

The same divisions that Stokes sees in his analysis of research policy show up in the daily routines of the people we have studied. Most of the inventors spend at least part of their time on each of the two questions, although their starting places and their patterns of attention differ. The pattern for the theoretician scientist corresponds to Bohr's quadrant; the pattern for engineers corresponds to Edison's quadrant; the pattern for designers falls in Pasteur's quadrant. Radical research—focusing particularly on the patterns of whole teams—fits best in Pasteur's quadrant.

Going to the Root in Pasteur's Quadrant

The approach of following a problem to the root in radical research can be understood as a dance within Pasteur's quadrant. It starts when an individual or a team tackles a difficult and important problem. The members of the team follow the problem wherever it takes them, bringing in additional disciplines as needed. This approach contrasts with single-discipline approaches, which abandon problems that seem outside their scope.

Radical research promotes a dance between the two questions. In this dance, the question "What is needed?" is reformulated every time a layer of the problem is peeled back to reveal a new fundamental challenge. When the deep nature of the challenge has been revealed, additional disciplines are added to the team. Their perspectives and new insights illuminate and reformulate the question "What is possible?"

An important operational difference between the linear model proposed by Bush and the quadrant model favored by Stokes is the handling of obstacles. The linear model illustrated in figure 2.6 features "applied research" as the next stage after "basic research" to solve particular problems. When an obstacle arises in applied research, the goal is to find some way to get around it. The team does not do experiments to discover new knowledge, because that is strictly the province of *basic* research.

In radical research, however, an obstacle becomes the new focus of attention and guides the process. When progress is stalled by an obstacle, this is a signal to dig deeper, introducing new disciplines as needed. In following a problem to its root, a new discipline might be added because it provides different and powerful methods, models, and metaphors. Sometimes the essence of the problem does not fit neatly into the concerns of any of the disciplines at hand. Often the insights that end up cracking the problem come from the space between disciplines.

Radical Research Teams in Pasteur's Quadrant

Engineers and scientists tend to follow the pattern that dominated their training. Engineers are trained to solve problems, placing them squarely in Edison's quadrant addressing what is needed. Scientists are trained to investigate phenomena, placing them squarely in Bohr's quadrant, creating new knowledge that addresses what is possible. Even though its engineers are in Edison's quadrant and its scientists are in Bohr's quadrant, a team doing radical research collectively acts in Pasteur's quadrant. This enhances the chances for breakthroughs.

The team pattern for radical research is as follows:

- *Engineering culture*. Focus on an important problem.
  This is why finding a solution has value.
  The problem pulls people together.
  It provides challenge and focus.
- *Science culture*. When you hit a barrier,
  analyze the barrier,
  create new knowledge at the barrier,
  and
  pull in more disciplines as needed.
- *Radical research culture*. Sustain commitment to follow the problem wherever it takes you.

In the case of the laser printer, there was a team made up of engineers and scientists. The common goal—creating a laser printer—was firmly established and unified the team. Several barriers were encountered in creating the laser printer itself, ranging from speed and timing issues to challenges in creating the flexible fonts. New knowledge was created at every barrier.[19] The breakthroughs ultimately had many unforeseen applications. Xerox founded a laser company to explore applications outside its core business.

Breakthroughs along the Road Less Traveled

If the problems of the future required only small incremental improvements on the well-worn paths of today's technology, there would be little need for basic research. If there were unlimited resources and no sense of urgency, there would be little reason for science to attend to society's real problems. However, this is not how things are. Society values breakthroughs because the future is full of surprises. Innovators pay attention to needs not only because needs inspire them but also because resources *are* precious.

Radical research requires a culture of innovation that can support open investigations into the underlying nature of problems and multidisciplinary collaboration. These cultural properties are not present in most companies or in most research institutions. This approach requires organizations to be flexible, because the people who start a search for a solution may discover that the real essence of the problem is different from what they expected and that different kinds of expertise are needed to address it.

Breakthroughs require nimbleness. Following a problem to its roots can take inventors off the road most traveled. Addressing the two questions involves learning how to dance.

3

Invention: Seeing Differently

I used to think "How could jazz musicians pick notes out of thin air?" I had no idea of the knowledge it took. It was like magic to me at the time.
—Calvin Hill, in *Thinking in Jazz* by Paul F. Berliner

When people hear about a breakthrough, they are surprised and often ask "How did they think of that?" A breakthrough can be surprising either because it works in a surprising new way or because it satisfies a surprising new need. In either case, the inventor saw something that others missed.

Inventors see the world differently from most people.[1] This has to do with *how they notice and understand* rather than what they are presented with. Sometimes inventors find curiosities in ordinary things. These curiosities become the seeds of their inventions.

Seeds of Invention

George de Mestral was a Swiss inventor and amateur mountaineer. One day in 1948 he took his dog for a nature hike. Returning from the hike, his clothing and his dog were both covered with burrs—the rough prickly seed packages created by plants that cling to animal fur, enabling them to hitch a ride to new places to grow.

De Mestral was curious. He took his clothing and the burrs to a microscope and observed that the burrs had thousands of tiny hooks that clung to tiny loops in the fabric of his clothing. He was fascinated by how the burrs clung to the cloth. Inspired to create a new kind of fastener that worked the same way, de Mestral collaborated with a weaver from a textile plant in France. They created two materials: a nylon cloth with tiny hooks and another with tiny loops. Patented in 1955, Velcro is now a major product used around the world.

Before de Mestral, millions of people had encountered burrs. Thousands of field biologists with microscopes have had the means to see the natural arrangement of hooks and loops. Most people simply remove the burrs with annoyance but without curiosity and without seeing a possibility for invention. George de Mestral found an invention where other people found just burrs. His experiences prepared him and enabled him to see burrs differently, as seeds for invention.

Four Colors of Invention and Discovery

There is not one right way to invent. Inventors find their creative inspiration in different ways, leading from their different strengths. But there do seem to be four main approaches[2] to breakthroughs, often used in combination. The four approaches can be combined, like the colors on a palette. We call the four main approaches *theory-driven, data-driven, method-driven*, and *need-driven* (figure 3.1)

Although many inventors and scientists use a combination of approaches, they often have a dominant one. Edison's main approach was need-driven; Einstein's was theory-driven. Mendel's was data-driven; Galileo's was method-driven. Pasteur famously moved between theory-driven and need-driven approaches.

De Mestral used the data-driven approach of invention when he noticed in the field that the burrs had a powerful sticking property. They stuck to cloth but not to his skin or boots. They could be removed but could stick on again. The data-driven approach notices patterns and anomalies in data. De Mestral wondered "What makes burrs cling so well?" He used the method-driven approach of invention when he obtained a microscope and looked at the burrs through it. The microscope enabled him to see the interaction between the hooks of the burr and the loops in the cloth of his clothing. With magnification, de Mestral could see how burrs worked. He used the need-driven approach when he realized that this sticking and re-sticking quality of the burrs could provide a new kind of fastener. He was inspired to invent a new technology that could replace zippers for many applications.

Figure 3.1
Four ways of inspiring discovery and invention. In the *theory-driven* approach, a mental model or theory provides a way of thinking that leads to insight and invention. In the *data-driven* approach, an anomaly in data reveals a surprising possibility. In the *method-driven* approach, instrumentation enables previously unknown observations, discoveries, and invention. The *need-driven* approach identifies problems and seeks solutions.

Perhaps the only approach missing in de Mestral's initial insight was the theory-driven one. He needed theory later to find a way to create burrs recreated artificially, imitating nature's engineering. The theory of manufacturing polymers was used in the synthesis of nylon materials to create hooks and loops.

George de Mestral prepared his mind to see an opportunity in nature's "engineering." At the age of 12 he designed and patented a toy airplane. Later he worked in the machine shop of a Swiss engineering company. De Mestral also knew about the frustrations of using zippers, such as when they get stuck and how difficult they are to use when your fingers are cold. In this way, his experiences on cold mountain hikes prepared his mind for recognizing what was needed. In short, de Mestral developed habits for thinking like an inventor. He cultivated ways of seeing differently.

There are two questions we might ask about the four approaches to invention: Are the categories of approaches universal? Are they meaningful and illuminating? On the universality issue, we note that all of the invention breakthrough examples that we have encountered either in famous stories of invention or from the interviews for this book follow one of the four approaches or a combination of them. Three of the approaches—starting from method-driven to data-driven to theory-driven—follow a path of cognitive engagement with the world and "information processing." Method-driven is about how information is gathered; data-driven is about how it is analyzed for patterns; theory-driven is about how it is understood and interpreted. In this way, these three methods correspond to three stages in gathering and processing information to create knowledge. The need-driven method relates to purpose—the sense that invention is not just about curiosity but also about making a difference in the world. In this way, the need-driven approach parallels the "What is needed?" question of the previous chapter and the other three approaches are sub-parts to answering "What is possible?" Seeing how inventors use these approaches may help budding inventors to be more deliberate and conscious of their inventive processes.

Stories

Theory-Driven Invention
Albert Einstein exemplifies the theory-driven approach. Einstein would imagine what could be seen by a rocket ship traveling at nearly the speed

of light, or what must happen to two photons traveling past each other in opposite directions. His great mastery of the thought experiment enabled him to discern deep contradictions in the dominant and competing theories of light: the particle and wave models. These thought experiments led him to relativity theory as a new synthesis of thinking and to testable experimental predictions.

Einstein is often quoted as having said "I am enough of an artist to draw freely upon my imagination. Imagination is more important than knowledge. Knowledge is limited. Imagination encircles the world."[3] Like Einstein, inventors and discoverers employing a theory-driven approach use an advantaged means such as thought experiments to aid their understanding.

Self-Configuring Robots   The traditional image of a robot is as an oversized and clumsy tin man. Industrial robots look nothing like this. Robots used in manufacturing cars tend to be built on fixed stands. They have cameras for eyes and mechanical arms with special attachments for hoisting parts and bolting or welding them into assemblies. Robot toys for kids are often much more versatile, such as the Transformer toys that change their shapes in order to exercise new capabilities.

In the late 1990s, robots closer to the Transformer ideal emerged from the modular robotics lab at PARC and found their way onto the sidewalks and gardens outside the building. To a significant degree, these robots could change their shapes and ways of moving. Sometimes they resembled a snake slithering along the ground, down the stairs, or through a tube. They climbed a fence with grasping hooks extended from their modules. They rode a tricycle with "limbs" acting like the feet of a young child, pumping the pedals. They could also curl up into a loop, and roll on the ground like a detached tank tread.

In the future, reconfigurable robots may be used in search and rescue operations, where they could slither inside collapsed buildings to reach people trapped in dangerous quarters, possibly bringing in supplies or using mechanical strength to create openings.

Mark Yim,[4] head of the modular robotics group at PARC:

My work on modular robots originally came from my thesis work. I was in the Mechanical Engineering Department at Stanford, but I was working under Jean-Claude Latombe in computer science. I was looking for a thesis topic. Mark Raibert had become well known at that time for building a one-legged hopping robot, and then two-legged and four-legged versions. I was also thinking about locomotion and thought that a cart wheeling robot would be interesting.

So I began thinking about how I could design a cart wheeling robot. It probably needed a bunch of legs. My design sense was that every leg should look the same so that I could design just one leg instead of a bunch of different legs. Next I realized that a leg has several joints. Could I design a leg so that every joint was the same? Then I realized that the robot did not need to be in the shape of a cart wheeling robot. I was explaining these ideas to my wife, puzzling about how to arrange it so that the joints would work together. She asked me "Why not have the joints just arrange themselves?" Everything fell out from there.

Before this work, robots were built from many special purpose parts. For the most part, every joint, every rod, and every limb on a robot was different and had to be designed separately. Yim wondered if he could build a complex robot without so much design work. The theory-driven core of Yim's invention was based on a simple question: Could a robot be built from a bunch of parts where all of the parts were the same? This idea is reflected in the name *modular robotics*. The modules provide a substrate or technology for configuring robots of many different functions and shapes—rather than building a different robot for each purpose. Yim:

A few years after I graduated, I took up the modular robotics theme again at PARC.[5] The first project was what we called Polybot. These were chains of robotic modules that you could control and manipulate together. More recently we started to work on what we call lattice re-configurable robots. These are more like Lego bricks that can reconfigure themselves. The core of the work is in designing algorithms that allow the modules to arrange themselves in different shapes and carry out different functions. A lot of computer scientists are getting interested in that type of re-configuration because it is easier to represent in a computer, yet it is still very rich in interesting problems. Imagine 160 or so modules arranged one after another like a "tape" of modules, or perhaps some other structure.

This use of programmable modules is the key theoretical shift that drives modular robotics. This kind of modularity translates what would have previously been a mechanical design problem into a configuration problem. It moves much of the experimentation from hardware experimentation to software experimentation. One of the things that I find really interesting is where hardware design is very tightly coupled with computation and software. There are often tradeoffs between what you can do on the mechanical electrical side and what you can do with software. The old paradigm for developing robots was that engineers developed the hardware or mechanical parts and then the software people figured out how to program it. The modular robotics design practice is a lot more fluid. In modular robotics you rearrange the same modules and program them for new functions.

Now we are doing something a little bit more applied, delivering a system to a customer. We are finding that more interesting at the moment. My long-term philosophy has been to do what's fun, which is pretty much "building cool stuff." That has worked out well for me probably because what I think is fun is often what other people find exciting too.

Research on modular robotics has continued at PARC. It is part of a larger research theme called "Smart Matter," which is concerned with deep tradeoffs between mechanical and computational aspects of materials and systems. The research in modular robotics was not initially started to solve a recognized need in the world. It grew out of a theoretical insight into design—essentially, playing with the idea of using many copies of the same module to build a larger robot. As the technology has matured, the project's focus has been shifting from developing radical new concepts to using the technology in designing new product architectures for Xerox. In this way, the approach has started to shift from theory-driven to need-driven.

Data-Driven Invention

Well-known researchers who used a data-driven approach include Alexander Fleming and Gregor Mendel. Fleming discovered the antibiotic effects of penicillin by noticing how a rogue colony of mold stopped staphylococci from growing on a culture dish. His path to discovery employed very careful observation. Another scientist might easily have just thrown away the anomalous culture dish, but Fleming noticed that it was unusual. He recognized its unusual property and used it experimentally to treat infections. In contrast to Edison, Fleming did not start out deliberately looking for a solution to a problem. He was known to have a very playful frame of mind. His playful curiosity and attention to data led him to the accidental *discovery* of penicillin.

Gregor Mendel's discovery was also due to his careful attention and immersion in data. Mendel discovered the rules of heredity for breeding plants. Mendel immersed himself in the data that he collected by systematically hybridizing sweet peas having different traits and keeping careful records of the results. His practice and data analysis led him to insights into sexual genetics and to guidelines for purposeful breeding. Although Mendel and Fleming both immersed themselves in their data, their styles differed. Where Fleming took advantage of a chance accident, Mendel explored breeding combinations systematically.

In data-driven invention an inventor notices something *anomalous* or *surprising* in the data and this leads to an idea. In his book *The Search*

Figure 3.2
A collection of early mice. Courtesy of Palo Alto Research Center.

for Solutions, Horace Judson quotes Lewis Thomas on what he notices in a laboratory on the verge of discovery:

One way to tell when something important is going on is by *laughter*. It seems to me that whenever I have been around a laboratory at a time when something very interesting has happened, it has at first seemed to be quite funny. There's laughter connected with the surprise—it *does* look funny. And whenever you hear laughter, and somebody saying "But that's *preposterous!*"—you can tell that things are going well and that something probably worth looking at has begun to happen in the lab. (p. 69)

Some kinds of patterns are hard to notice. Sometimes the patterns are of a different kind than was expected and are not recognized. Sometimes a discovery starts by just noticing something subtly different about a particular sample or experiment. Success in the data-driven approach comes from being *immersed* in the data and paying attention.

Noticing Inefficiencies in Hand Motion   The mouse is the most common pointing device for desktop computers, but many users of laptop computers prefer a pointing device that is integrated with the keyboard. They need to set up their computers often and use them in tight quarters. One of the most widely used pointing devices for keyboards is the TrackPoint, the eraser-like nub that appears between the G, H, and B keys on IBM ThinkPad laptop computers and many others. The TrackPoint is largely the invention of Ted Selker, now a professor in the Media Lab at the Massachusetts Institute of Technology.

Ted Selker sees invention mainly as problem solving. He considers himself to be an applied scientist and an engineer. Selker was working at Stanford, and Atari and was thinking about pointing devices. He had

Figure 3.3
A TrackPoint pointing device on the keyboard of a laptop computer. This device was developed by IBM.

read a classic book[6] on user interface design, where he had seen an analysis of the use of the mouse. Earlier, Selker had experimented with pads similar to those on current keyboards. We asked Selker to describe how he invented the TrackPoint[7]:

In Card's book there was a GOMS-style[8] analysis of how long it takes your hand to get from the mouse to the keyboard and from the keyboard to the mouse. At the time I had this big fascination with knee bars. A knee bar is that bar you use on a sewing machine to control its sewing speed. The data showed that a knee bar is better than a mouse for the first fifteen minutes of use. I was wondering "Why would the knee be any good for control? The knee has almost no sensory or motor control compared to the fingers or the hands. So why would the knee bar be any good?"

It's good because you don't have to take your hands off the keyboard! I found that it takes $1\frac{3}{4}$ second to move your hand from the keyboard to the mouse. One and three quarters second is a long time. You also have to move your eyes and look for the mouse. I realized that this movement could be 25 percent of the speed of your typing in some editing tasks, going back and forth between the mouse and the keyboard.

The $1\frac{3}{4}$ second that it took to grab the mouse caught Selker's attention. He realized that someone doing an editing task would spend that time every time he needed to grab the mouse, such as to scroll to a new place in the text or to select a menu item.

Progress for the TrackPoint was not straight and smooth. Selker had to stop working on pointing devices for 2 or 3 years when he first joined IBM. He resumed this research when another colleague joined him, and they began building and experimenting with prototypes:

We started building and inventing sensors. The first one was a bizarre sensor I designed made out of resistive foam for packaging chips.

We did some calculations and found that getting your finger to the TrackPoint took time, but that we could save a second over the mouse time. This was not the full second and three quarters, but there were a few places on the keyboard where you could locate the TrackPoint on the keyboard and get your finger there a second faster—or at least 0.9 second faster, which is an awful lot. That was very exciting. We tried the TrackPoint in at least five different places on the keyboard.

An insight from data causes an inventor to see differently. These seconds, fascinating to Selker, had been overlooked by others. For Ted they jumped out! "This is interesting!" There had to be a better approach that would save the "mouse-reaching" time. This reminded him about some experiments with a knee bar. The knee bar was good because you didn't have to move your hand away from the keyboard to reach for a controller. By building a pointing device into a keyboard, perhaps reaching for the pointer could be more like moving a finger in touch typing. Focusing on the data opened up a new possibility for invention.

Method-Driven Invention

Galileo used the method-driven approach. Using a telescope, Galileo was able to see the moons of Jupiter, which are not visible to the naked eye. Similarly, the microscope led Robert Hooke, Antony van Leeuwenhoek, and others to the discovery of bacteria and the first observations of living cells. Similar observations enabled Louis Pasteur and others to understand the role of microorganisms such as infectious bacteria and yeast. This led to the development of better ways of sterilization and reliable ways for making wine.

New instruments do not always lead to acceptance of new ideas. Galileo's difficulties with the Roman Catholic Church on the meaning of his observations almost cost him his life. However, the operation of telescopes was not greatly questioned, since telescopes could be tried and tested on distant terrestrial sights. In other words, you could just travel down the road and confirm that the image in the telescope was accurate. People were much more suspicious about the images produced by microscopes. The new worlds seen by Hooke and others were treated with suspicion for many decades, perhaps since the findings could not otherwise be confirmed.

The method-driven approach also applies to experimental techniques. Examples include techniques for combinatorial genetics and combinatorial chemistry. These combinatorial methods enable researchers to carry out thousands of experiments at once.

Improving on Windows   Stuart Card is a senior research fellow at PARC and head of its User Interface Research group. The work of his research group is multi-disciplinary. Cross-talk in his group meetings always includes intersecting themes from psychology, computer science, and human-computer interfaces. Such cross-disciplinary research led to an invention for supporting people who are constantly multi-tasking on their computers.

Window-based systems had been used at PARC since the mid 1970s and had always had some usability problems. In this story, Card recalls the invention of the "Rooms" system in the early 1980s. Rooms fixed many of the problems with earlier window systems. At the same time, PC-based windows systems from Apple and Microsoft were beginning to appear on the market, with all the problems of the older windows systems. Card helped XSoft (a Xerox subsidiary) develop several Rooms-like products for the early PC market. The Rooms system was in many ways easier to use and more powerful than the systems that are widely available today in commercial window systems. Card told us what motivated him to start working on Rooms[9]:

It bothered me that windows had been very successful on workstations and very unsuccessful on PCs. What was the reason for that?

I was trying to understand the effectiveness of window systems. A couple of people had tried to do experiments on windows. Getting meaningful results was difficult because the measurements depended on the combinations of things that subjects were doing. The researchers could not make much headway in understanding what was happening.

There was also a problem with windows—high overhead. You may remember that Star[10] had *tiled* windows[11] because the designers thought that *overlapped* windows had too much overhead.

Figure 3.4
A Xerox computer running Rooms software. Each region of the display is an overview of a set of windows. Courtesy of Palo Alto Research Center.

I wanted to understand the overhead problem. First I looked at video tapes of people using windows. Then I got instrumentation with the help of Richard Burton and Austin Henderson. That gave data about window locations. I did an experiment in which I instrumented every window. But first I looked at the video-tapes and plotted the data in my notebook, noticing how many pixels were being used and what windows were overlapping.

One of the patterns that was really interesting—as somebody is working—is that there get to be more and more windows. When they finish a task, they clean up the windows. So if you look at the pixel loading, the windows build up and then go away as users finish and clean up.

I started thinking of an analogy to how programs use memory locations in main memory. I tested for locality of reference.[12] And I found locality of reference with the help of this program that Austin and Richard created. Then I remembered a *Computing Survey* article on virtual memory by Peter Denning. It showed that the average access time for a virtual memory operating system was very close to what it was if the memory was really that big. Astonishing!

In the computer operating systems of that time, a computer application (such as an email program) could have several windows. As a person worked on a task, the number of windows in use would increase and they would overlap. At the end of a task, one would clean up all the windows. Although this behavior was there for anyone to see, Card was the first to call it out and find it interesting as a cognitive psychologist. By collecting data from video tapes and internal program monitors, Card was able to see patterns of window usage in the behavior of his subjects. He could also see what was happening when the systems became frustrating and the activity began to break down.

As he worked to understand the data, Card employed a theory-driven approach. His computer science background kicked in. He remembered a widely read paper by Peter Denning about computer virtual memory and paging systems. Denning explained how such systems worked and what governed their performance. Computers can run programs larger than the size of their fast main memory by "swapping out" pages[13] of the program or data to a slower disk memory. When a program needs to refer to that memory page again, it selects a different page to copy out to the disk, copies the desired page back into the main memory, and then resumes. The event when a program asks to access a page not available in main memory is called a "page fault." A computer program tends to use a fixed set of pages for a relatively long time between page faults. When this is true, swapping overhead is low and the computer system runs almost as fast as if it had enough main memory for all of the pages.

This is where the theory-driven approach came in. Card saw an analogy of paging behavior to window usage. Since screen size is limited, people cover windows up to make room for new windows. By analogy, a limited-size display is like a limited-size main memory, and covered-up windows are like swapped-out pages. Just as computers have to swap in a memory page in order to use it, people have to pop a window to the top in order to use it. Stu called the event of bringing a covered window back to the top a "window fault." He decided to build on Denning's analysis of paging systems in order to design a better windows system:

We designed rooms using the abstraction of virtual memory systems. As we got started, Austin had the idea that we were going to create a giant workspace, which was called BigScreen. We were going to move all around in the landscape. The problem was it's so big. So we did an overview. Well, that doesn't scale, when you get enough windows on the screen.

So we tried putting signs around. For example, a "Mail" sign would indicate the region where all of the mail windows were. As the space got bigger, we had to make the signs relatively bigger in the overview. We made *really big signs*! And we had about a dozen techniques that we tried to use in order to make this overview so you could find your way around. Then we noticed that the windows clustered into little groups. The next step is you realize that if I just had a bloody pop-up menu, I could go to the mail section and I wouldn't have this gratuitous search through the space.

Card and Henderson experimented with dozens of windows in the big space. They created giant labels for the clusters of windows that people used together. At some point, they decided to separate the clusters into named groups of windows called "Rooms." You would go into a room to do a certain task. "This is my email room; this is my writing room for paper number 1; this is my programming room; and so on." The rooms metaphor led them to have "doors" between the rooms, overview maps of the rooms, and so on:

The first design was naive in several ways and ran into problems. Because I had this bit of theory, I could understand what I was seeing in those problems. You know, "Discovery favors the prepared mind." So the window faulting theory was the preparation. The *theory* gives you the eyes to see the behavior and what your problems are. This is what Herb Simon[14] was referring to when he advised "Always have a secret weapon."

Later when I started the information visualization research, Allen Newell[15] used to ask me "What's your secret weapon? What's your theory doing down below?" He was implying that if you didn't have a theory, you were no better than any other sod trying to invent stuff. [laughs]

The theory for Rooms grew out of an analogy to operating systems. Operating systems need to run big programs in small computer memories. Card made an analogy to people using big collections of windows on small displays. In operating systems, units of memory or pages of data and program are moved from disk memory to main memory when they are needed. In windows systems, windows are unburied from below other windows when they are needed. Card found himself looking for effective ways to manage screen space, and this led him to examine the techniques for managing memory.

Having a method of collecting new kinds of data allowed Card to see something that had gone unnoticed before. Having a theory enabled him to understand what to look for in the data and also guided his thinking toward a solution.

Need-Driven Invention

Thomas Edison exemplifies the need-driven approach, since his work was largely inspired by particular problems that he wanted to solve. In his effort to perfect the electric light bulb, Edison is legendary for his tireless testing of one material after another before finding a workable electrical filament. With the phonograph, the stock ticker, the carbon microphone, and many other inventions, he directed his attention and that of his associates toward creating things he felt the world needed.

Need-driven invention is when an inventor sees an unresolved need in the world. The need-driven approach may be the most common one. Sometimes need-driven invention takes the form of *gap analysis*, where current technology is almost good enough but there is still a gap before it can be useful. Gap analysis focuses effort on a gap where progress would enable a solution. Need-driven inventors are consumed by their search for solutions. Whenever they look around in the world, they search for clues or ideas that might contribute in some way to solving the problem at the back of their mind.

Dark Rooms and Dim Screens   The electric paper or "gyricon" project began in the 1970s when Nick Sheridon noticed people making their offices dark in order to use their computers. This was during the time of the Alto,[16] the ground-breaking personal computer for which PARC is famous. The problem that Sheridon saw was that the displays on the first

Alto computers were not good enough. Sheridon believed that a new kind of display was needed.

We asked Sheridon how he got started working on electric paper[17]:

I started working on the gyricon project at about the time that the Alto was introduced at PARC. The interesting thing is that if you went into people's offices and they were using an Alto—they had closed the blinds and the drapes, turned off the lights, and often closed the door. They made the room as dark as possible so that they could work on the computer.

The reason for keeping the room dark was that the video screen on the computer was of poor quality. There was poor contrast, reflections, and flicker of the screen. Some of the people found the displays very difficult to work with. I decided to try to develop a display that would work well in broad daylight. It should work as well at the beach as in a dark room. Like paper the display would work with reflected light.

I came up with two or three ideas and selected one, which I thought would be the easiest to do. That's where the gyricon idea came up. That started it.

The gyricon displays invented by Sheridon are micro-mechanical displays controlled by an electric field. There is a plastic sheet with a thin inner layer of microscopic balls resting in fluid-filled cells. The balls are typically black on one side and white on the other. Under the influence of an electric field, the balls can be selected and rotated to show either their black sides or their white sides. Once the balls are rotated to a position, they stay put until another charge is applied. The black side of a ball is like black ink on a page, and the white side is like white paper. Sheridon's invention makes the microstructure of electric paper become changeable so that it can display whatever information is required without backlighting. In effect, rotating a ball is like moving the "ink" between the front and back sides of the paper.

Unlike the images on a television tube, which must be repainted many times a second even if the image isn't changing, a gyricon display does not require refreshing. Furthermore, the display can be read easily in bright light. Because it is illuminated by ambient light, there is no need for a power-hungry backlight. Sheridon:

I worked on that technology for about a year. I published a couple of papers on it and took out a couple of patents, but I was asked to discontinue the work. That was about 1973. I was told that Xerox was not a display company and that we were not going to go into Alto production. I was encouraged to work on something like printing where the Japanese were really eating Xerox's lunch and see if I could come up with a better way of doing electronic marking.

The electronic paper concept evolved over a period of time. The first gyricon concepts were developed in the 1970s, but the real electric paper concept to imitate paper more closely came in the early 1990s. Paper is the perfect display. Anything else is a poor imitation and the closer that you can come to paper the better. At the time of the Alto I wasn't trying to recreate *all* of the properties of paper in a display. I wanted something that had the visibility qualities of paper, but I didn't care about weight and flexibility. When I came back to the project later, I wanted to make a paper substitute that had *all* the important properties of paper. Today, electric paper is a flexible material that you can carry around rolled up like a newspaper.

In the new electric paper project, I was motivated by watching people doing their e-mail. A lot of people print out their email or short papers on paper, read it, and then toss it out in five minutes. Electric paper could be used *and reused* in inkless electric printers. So there was basically an ecological concern. If you can reuse electric paper a million times, then it's a whole lot cheaper than paper. Also the printer would be cheaper than a printer that moves pigment around.

Several years of experimentation eventually yielded a working technology. The first commercial applications of electric paper are wireless, battery-powered, variable-data signs for use in retail stores. In the spring of 2001, Xerox PARC spun off a company, Gyricon Multimedia, to commercialize its electric paper technology. Sheridon is now that company's Director of Research.

Finding Good Music   Many observers of the information age have noticed that the amount of published work available to us—in such forms as articles, books, web pages, movies, and music—is growing exponentially. However, our human capabilities for absorbing, reading, or using these works are not increasing at all. As Nobel laureate Herbert Simon put it, an abundance of information leads to a scarcity of attention.

One way to learn what is good is to collect data on what others like. Suppose our computers gave everyone a way to indicate what they like and then gathered and pooled their opinions. We could use the data to focus our attention. A complication is that individuals have different tastes. Suppose that we could get sharper or more compatible advice by combining our judgments with those of *like-minded* people, sharing our similar judgments about (say) movies and music? Such selective sharing is the basic idea behind collaborative filtering.

One of the leading innovators in collaborative filtering was Pattie Maes, a professor in MIT's Media Lab. When we asked her how the idea of collaborative filtering got started, Maes replied as follows[18]:

My work is motivated by problems or opportunities that I see for doing things in a more efficient way, or in a different way than they are currently done. So I'm very interested in real world problems and inefficiencies.

Our research in collaborative filtering was motivated by the fact that I couldn't find music that I liked. When you went through a record store in those days, you couldn't listen to the records. You could see the covers but it was hard to know what the music was like. I didn't like the radio stations in Boston when I moved here from Brussels, and I felt there had to be a better way to find out about new music and keep informed about music that may be of interest. So the whole idea of using a computer to facilitate spreading information by word of mouth really came from my interest in finding music.

When I actually start a project, the exact problem may not be clear yet. I couldn't say "OK, I'm going to build this system or test this hypothesis." It's more of an intuition or a feeling that if I start on a particular approach, I will hit on something worth solving.

Maes's reflections on the origins of collaborative filtering[19] show how her first steps involved subtleties not evident in a simple "find a problem, then solve it" prescription for need-driven invention. The basic idea of collaborative filtering *arose* from the problem of finding music, but the key elements of the invention revealed themselves only after the work started. Maes:

Most of the work that we do is like this. We start with a half-baked idea which most people—especially critical people—would just shoot down right away or find uninteresting. But when we start working on it and start building, the ideas evolve. That's really the method that we use at the Media Lab. We start building systems as opposed to doing a *lot* of designing and thinking ahead of time. The project *evolves* into something that becomes more and more interesting. In the process of building something we often discover the interesting problems and the interesting things to write about and that leads to interesting discoveries.

This was true with the music recommendation system. Initially, it seemed like a very simple idea. People would tell the system what they like and then the system would tell them what music other people like with similar tastes. This resulted in a much bigger and more interesting project. We started introducing people to each other that had similar taste and developed many alternative methods for exploring a "taste space." There were many other projects that spun off from that very first idea and the first system that we built. It became a source of many interesting research projects.

Which comes first: starting the invention, or finding the problem? What is the first step: finding a crisp problem to solve, or starting work in an area in order to reveal what the interesting problems are? Teasing apart the inventive process in cases like this reveals some of the skills of inventors and some of the variations in the methods that they use.

Today there are several companies on the World Wide Web that use collaborative filtering to rate movies, music, and other products. One of the first commercial systems for collaborative filtering was Firefly, developed by Pattie Maes and others at MIT. Maes and her graduate students commercialized Firefly by creating the company Agents Inc., later renamed Firefly Networks, Inc. The company was purchased by Microsoft in the late 1990s.

For Nick Sheridon and Pattie Maes, having a need in mind changes the way that they see the world. Whenever they see something, they are always searching for a solution to the problems they have in mind. Knowing the need causes them to see differently and to look in places where they think they will find inspiration or answers.

Reflections

Summarizing the stories in this chapter, four main approaches govern how scientists and inventors are inspired in breakthroughs and discoveries. These approaches are like colors on a palette in that they can be mixed to form variations and combinations.

1. *Theory-driven invention.* A tag line for this approach might be "Eureka!" or "According to my theory . . ." This approach uses a theory, a model, or an analogy as a tool for thinking. These tools for thought provide advantages and shortcuts to insights leading to invention.

2. *Data-driven invention.* The tag line for this approach might be "That's strange!" An inventor notices an anomaly or something surprising in the data. Paying attention in this way creates an advantage, leading to insights and invention.

3. *Method-driven invention.* The tag line for this approach might be "Now I can see it!" Researchers have a new instrument that enables them to observe things not visible before.[20] The new instruments give them advantages in observation, leading to insights and invention.

4. *Need-driven invention.* The tag line for this approach is "Necessity is the mother of invention." An inventor learns about an unresolved need or problem in the world and searches for a way to satisfy it or solve it. This approach fosters invention because the problem rests at the back of the mind as an unresolved challenge. It becomes a backdrop for interpreting experiences all day long. In effect, an inventor thinks about whatever ideas or observations show up as elements of possible solutions.

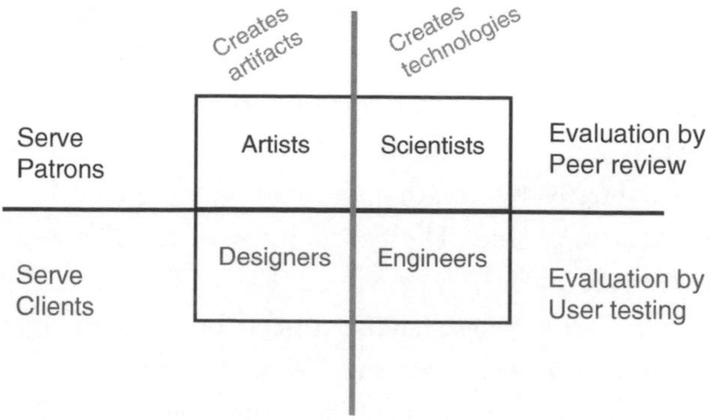

Figure 3.5
Relationships between four of the creative professions. This figure is based on a suggestion by Rich Gold.

Scientists, Engineers, Artists, and Designers

As this chapter describes them, the four ways of seeing differently have a distinctly *scientific* flavor. They correspond to steps in an experimental scientific process. The method-driven approach uses advantaged ways of collecting data; the data-driven approach involves being immersed in experimental data; the theory-driven approach draws insights from a theoretical framework.

However, "scientists" are not the only people who are creative and invent things. What about engineers and designers? A lot of breakthrough inventions rely on fundamental advances in technology, so it is not surprising that scientists contributed to most of the big breakthroughs of the 20th century.

Figure 3.5 characterizes some of the relationships between four of the creative professions.[21] The rows and columns capture signal broad and important similarities and differences among these professions.

The left column—comprising artists and designers—represents creative professions that create artifacts. We might also put architects—designers who work on buildings and other large public structures—in that column. The right column, comprising scientists and engineers, represents creative professions that create technologies. This distinction is not perfect, of course. For example, engineers also create artifacts, and some scientists create only theories, which are not properly artifacts.

However, in the context of breakthrough innovation, these characterizations will at least help to frame some issues about invention.

In many cases, designers see artists as practicing a purer form of their work and having more freedom. In some cases, engineers see scientists in an analogous way. Underlying this, however, are some other salient differences between the top row and the bottom row of the figure.

In broad historical terms, the professions in the top row—artists and scientists—have had their work funded under a patron system. In a patron system, a wealthy patron—a government, a large corporation, a university, a wealthy individual—provides funding for the support of the professional and the work but does not try to manage it directly. The evaluation method for the work is based on a kind of peer review. In modern science, peer review plays heavily in federal funding agencies and in the vetting process for scientific publications.

In broad terms, the professions in the bottom row—designers and engineers—have their work funded under a client system. In a client system, work is funded by the period or by the piece to satisfy the needs of a client. Evaluation of the work is done by user testing, often by the client.

There are many deep similarities and affinities among all the creative professions. Although all of these professions are represented to some extent in many large research centers, the population of these centers is usually biased such that scientists form the dominant population, followed by engineers, and designers with artists usually making up a very small minority.

How do the four approaches to invention work out when we consider all these creative professions? Clearly artists and designers are known and honored for their powerful ways of "seeing differently," but do they invent differently from scientists?

We have not gathered enough examples of designers at work to offer confident findings about their innovation methods.[22] It is worth noting, however, that design schools often focus on rapid prototyping, building prototype versions of a product in order to experience them directly. Experience with the prototype becomes the basis for detailed observation and inspiration for the next round of design. In terms of the four approaches to invention in this chapter, this approach might be characterized as a combination of method-driven and data-driven. It is like the method-driven approach in that powerful observations are advantaged by having the device itself. It is like data-driven in that inspiration arises in the process of making careful observations of the prototype. Since both designers and engineers tend work with clients, their professions require them to develop skills for need-driven innovation where their clients identify needs.

The professions of designer and engineer also have in common the practice of building things from existing components, rather than creating new technologies. One way to advantage invention by designers and engineers is to give them new technologies that are created by the scientific staff. When engineers and designers practice need-driven invention with a background of privileged technologies, this can lead to breakthroughs in directions not anticipated by the original creators of the technologies.

4

Innovation: The Long Road

But you didn't stop.
You knew what you had to do,
though the wind pried
with its stiff fingers
at the very foundations,
though their melancholy
was terrible.
—from "The Journey," by Mary Oliver

Invention is having an idea; innovation is the other 99 percent of the work. From a business perspective, innovation is "taking new technology and realizing its economic potential."[1]

The creative insights of invention can happen in a flash. In contrast, innovation can take years (figure 4.1). To succeed as an innovation, an invention must prove workable. It must solve a problem worth solving. It must be better than competing inventions for the same purpose. Many hands and many skills are usually needed, each contributing, testing, and refining the products. Everyone involved must work together with common purpose.

When an invention is developed across multiple organizations, each of which takes it further, the process is called *technology transfer*. This process is not as simple as "throwing a technology over the wall."[2] The secret of successful technology transfer is in managing relationships. Failure to develop joint goals and commitments and to follow through at each stage will result in technologies and products unnecessarily being dropped.

When product success depends on more than one invention, an entire system of innovations must be developed. For example, the deployment

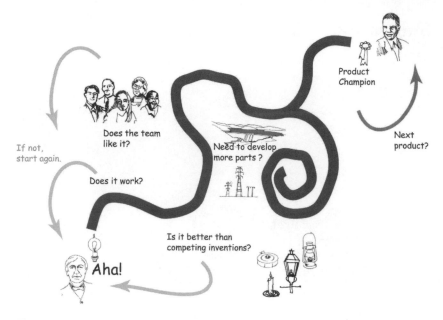

Figure 4.1
There are many checkpoints and detours on the long road from invention to innovation.

of electric lighting involved not only light bulbs but also outlets and switches, neighborhood and regional electrical distribution, standards for wiring, electrical generation, and so on. Difficulties in innovation are what Thomas Edison had in mind when he characterized genius as 1 percent inspiration and 99 percent perspiration.

There are many roads to innovation, each with its particular challenges. One inventor may encounter difficulties with competing products; another may face a series of daunting technical problems; another may work hard for a market that materializes very slowly or depends on establishing industry standards. Road maps are different because the challenges are different. Innovation is getting an invention into the world, doing whatever it takes. Here are some cases that show what it has taken.

Stories

Solving All of the Problems
In 1946 twenty people met in a burned-out department store in downtown Tokyo to form a new company—the Tokyo Telecommunications

Engineering Corporation. One of the founders, Masaru Ibuka, wanted to create a product that would be of general use to the public. The tape recorder looked like a good consumer product for the new Japanese company.

The wire recorder[3] had been invented years earlier in Germany but had many limitations. In the United States, tape recorders were made by Ampex and magnetic recording tapes were made by the Minnesota Mining and Manufacturing Corporation (now known as 3 M). In Japan, however, the knowledge about how to do this was missing. The electronics and mechanical design seemed within reach, but no one in Japan had experience making recording tape. Nobutoshi Kihara was a young engineer with the new company and a former student of Ibuka. He was on the front line for solving many of the technical problems. Kihara[4]:

Mr. Ibuka offered me the challenge of developing a tape recorder in July, 1949. At that time tape recorders were mainly for business use, but we believed that a tape recorder would be useful for entertainment in homes. All men have dreams that inspire them. Our dream was to record musical performances. As an engineer, I believed that it was very important to be able to record conversations and music.

Our first problem was in making a workable recording tape. The only information we had at first was a reference book that said that the AEG company in Germany had invented a tape recorder that used a plastic tape coated with a magnetic material. That was the only thing we could find about tape recorders in the book. We did not know what kind of materials would be suitable for making the tape base or for the magnetic coating. At first I tried to grind a magnet to make powder. However this did not work. The magnetic properties of the powder were too strong and you could not record or erase a tape made this way.

In school I asked for advice from other researchers and continued to search for the right material. Eventually I found a reference by Kotaro Honda about a promising chemical called oxalic ferrite. After the war it was very difficult to find this chemical. Akio Morita offered to help obtain it and we went off to search in the pharmaceutical district. We searched for a long time. We finally found the only pharmaceutical dealer who carried it. We bought two reagent bottles of the material and brought it back for further experiments.

The oxalic ferrite needed to be treated. We borrowed a frying pan to roast the powder. The heat-treatment was difficult. If I was inattentive for a moment when heating the oxalic ferrite in the frying pan, it burned immediately and became colcothar. It was important to stop the advance of the chemical reaction when the material turned blackish brown. The brown was ferric oxide and the black was ferrous tetroxide.

The next problem was to find a way to apply the magnetic powder smoothly to a tape. We kept working by trial and error. I tried various adhesives, such as a solution from cooked rice and Arabia paste. However, the results were not satisfactory. Then, I tried using a spray gun which painters had for painting the interiors of the new office buildings under construction in those days. The results were good.

Many other problems surfaced, such as producing fine enough powder and developing the capstans for reel-to-reel transport and keeping the right tension on the tape. There were problems regulating speed with the weak induction motors and the available natural rubbers that stretched and snapped. Kihara reflected on what is important in developing skills as an inventor:

I think it is important to develop ideas experientially. Know-how is not accumulated only by reading other people's books. For me, it has always been important to work with things directly with one's own hands, such as "wiring by oneself," "grasping a soldering bit," and even "developing an algorithm or analyzing by computer."

Many further challenges lay ahead in developing and marketing the tape recorder. They had to understand the market. When the first machines proved too expensive to sell for home use, they found that they could sell them to courts to record proceedings. Later they worked with music schools. The tape recorder was the first commercially successful product of a company known today as Sony.

Evolving a Series of Products

If you have ever used a web-based search engine, a word processor, or a language-based hand-held device, there is a good chance that you have used the inventions of Ron Kaplan and his group. Kaplan initially created the technology known as finite-state morphology—an efficient way of describing computationally which combinations of letters or symbols correspond to valid words. The technology had immediate applications in the language-based hand-held devices such as spelling checkers, dictionaries, and translation aids that were developed in the 1980s. Xerox PARC created a partnership with Microlytics Incorporated to commercialize the technology for these markets. Finite-state grammars were thought to be the easy part of a long range research program into computational linguistics[5] and natural language understanding systems.

We asked Kaplan how he got started developing products based on the morphological analyzer.[6] He replied:

How do you take a large finite-state transducer[7]—we were up to about a hundred thousand states in the dictionary—and how do you represent that in a very small amount of space?[8] This was a new question. We always thought that speed was the most important parameter, but in our first practical system, we found out that it was space that mattered, and not time.[9]

We got a market request from Microlytics that said "We really need the recognizer to fit in small space." This completely changed our thinking. We realized we could do this tradeoff in a different way. The system didn't have to be all that fast. It had to be small.

Space was crucial because space was what you paid for. Space went into the UMC[10] and "time" was just the user experience and irrelevant to the purchase price. For these hand-helds time was plentiful. From a computer's perspective, there was nearly an "infinite" amount of time to respond, given the slow one-finger typing of the user.

Kaplan took a year's leave from his research responsibilities at Xerox PARC to develop products with Microlytics. After developing finite-state dictionaries for hand-held devices, they were asked to create spelling dictionaries for the first personal computers. The requirement for small space applied for this too because of the need to fit huge spelling dictionaries on relatively small floppy disks:

The difference between being able to fit all of the data that you needed on one floppy to do a spelling checker or thesaurus and two floppies on an 8086[11] was the difference between being able to sell it or not. People would not do spelling checking moving one floppy in and out. So 300 K or whatever it was in those days was the deciding limit for products on a PC.

Technology keeps changing and providing new opportunities for applications of finite-state morphology. As the Internet began to blossom in the 1990s, the computer capabilities had advanced a lot but the need for spelling checkers endured. Kaplan found that another generation of systems needed to be developed with different requirements. In this case, saving space was no longer the issue. Speed was the issue. Xerox PARC formed a spinoff company called Inxight to commercialize the new versions of the technology. Large dictionaries had been carefully developed. However, they were not "programmed" directly into Inxight products. Rather, the idea was to modify the linguistic compilers so that lookup became much faster, even though the dictionaries required more computer memory:

At one point it was at 74 words per second. Then I got it up to 20 or 30 thousand. That's what enabled Inxight to compete against special purpose algorithms that other people had developed, which did not provide the architecture to carry across languages.

Kaplan and his team faced repeated challenges as the underlying linguistic technology was developed and then re-optimized for several generations of computing equipment. Although they had to adopt the compilers for multiple languages, they did not have to code the dictionaries differently by hand. They used a compiler to translate the dictionaries differently for different applications, optimizing space for hand-held devices and optimizing time for Internet applications. This made it possible to keep and maintain one dictionary and to repurpose it for different applications.

Enrolling a Development Team

Ted Selker conducted human-factors studies to measure people's performance using the TrackPoint.[12] For example, in a series of experiments in his laboratory, Selker became convinced that the stiffer the TrackPoint, the better the human performance. However, performance was not the only thing that mattered. Selker found this out as he worked with the product-development teams[13]:

The TrackPoint started out initially as a kind of bootleg project. The biggest problems at the beginning were non-linearity and non-monotonicity in the sensors. The sensors sucked. I went to my manager and asked if I could spend $500 on a sensor from Measurement Systems Corporation. He looked very puzzled and discouraged and said "Buy two." It was funny because up to that point this was an almost unauthorized project. I had taken a broken microscope that I had fixed, and an oscilloscope that was in the hall that somebody had given up using. We started in this very low grade way where my friend Joe Rutledge started testing the waters of his collaboration with me and testing the waters of his interest in electronics and so on. By doing it that way, a progressive commitment occurred.

We often had theories that were debunked by other people. For example one theory was "the device is faster if it's absolutely stiff and has a very stiff connection to your finger."

When we were about to put it into a product, one of the managers pressed his finger onto it and said "See that dent. I don't want to see a dent like that on any customer's finger." We said "But it's like a BMW, not a Cadillac. It has tight steering and stiff suspension. You've got to have that taut physical connection."

But it was no use. The manager did not like the dent. So we went down to the lab.

Another issue that Selker had not grappled with was that the purchasing decision for buying a computer with a TrackPoint is usually made by a person who has never used one. Since the early version of the

TrackPoint took more than one minute to master, the initial bad impressions would mean that the computer might not get purchased at all. The perception of the first-time user had to be addressed:

For everyone *except the absolute novice*, the TrackPoint was at least 20 percent faster at a mixed typing and editing task. And after 40 minutes, it was even faster for continuous selection than a mouse.

But not everyone was satisfied. I attended a business meeting where one guy focused his whole talk on the first 1½ minutes of experience for a first-time user. I countered with "I went to Akihabara[14] yesterday and I watched a person playing with a computer with a trackball and they were trying to decide which computer to buy. I clocked them at a half hour comparing computers and trying to make their purchase decision." He said "That's not normal. People don't usually spend that much time deciding. You've got to solve that finger-pressure problem and you've got to fix that initial experience for the first-time user." Satisfying his criteria was standing between me and getting the TrackPoint into a product.

His suggestions and requirements led to improvements. One of the interesting things he said was "You have to work tightly with Yomada-san." Now Yomada had just spent the last six months working tightly with manufacturers of trackballs and was completely oriented toward his relationships with those companies and his trackball. I had to overcome his objections and I had four days to do it.

We bought 40 different kinds of materials Quartz, cork, foam, sorbothane,[15] sandpaper, and so on. Yomada spent a lot of time with us in the lab. I finally created an air-suspension top that smushed around your finger rather than bit into it. We found out that by having the finger move a small amount with an air suspension and a gooey top, we could get a 15 percent improvement in pointing performance for the novice user.

That was wonderful because Yamada was the main objector to this technology. It helps to include people in the process when you are solving their problems. I believe that including Yamada was the key to succeeding. What the Japanese really wanted, what they always seem to want, and what I really enjoy about their approach to innovation and product technology, is that if we're going to have a solution we're going to have a *team* consensus. *All* members of the team have to contribute. If you don't include everyone, they don't believe you. I learned more from that process than almost every other part of the experience. When adversaries were willing to engage in creating a solution, we could make more inventive progress than when adversaries just had to be convinced, and would only listen to "the facts, ma'am."

There are a lot of good things about progressive commitment. When we first started working on the idea, I thought it was a six week project. Many years went by before we got to the part of the story where we worked with Yamada. Overall, it was a ten-year project. I would have never gotten involved in it if I had thought it was going to consume so much of my time.

With TrackPoint, as with many technologies, the time from "Aha!" to product was several years. In some cases the same technology is used for different products, but it has to evolve differently for different products. That is what happened in the next story.

Beating Competing Inventions

The first computer mouse was invented by Doug Engelbart and his colleagues at the Stanford Research Institute (now SRI International). During the mid 1970s, Bill English and several other members of Engelbart's group left SRI and came to Xerox PARC, where the first personal computer, the Alto, was created. Over the next two decades, many people at various companies made contributions to generations of "mouse technology"—the first optical mouse, the first optical mouse on an arbitrary surface, the first mouse at Apple, the first wireless mouse, and so on.

Bill English was working on versions of the mouse for the Alto and asked Stuart Card to help him to evaluate the mouse and to compare it to other alternatives for pointing devices. Versions of the mouse had been demonstrated, but questioned remained. Was the mouse the best pointing device for personal computers? What qualities for the mouse were important in optimizing human performance?

Stuart Card is a psychologist and computer scientist who is interested in understanding how people interact with information. If someone asks him about his background in psychology, he is quick to point out that his work does not have much to do with clinical psychology. Rather, his focus is on human performance for tasks that involve perception, hand-eye coordination, and information processing.

Before the mouse became established, there were many competing devices that could have become the main pointing device for personal computers. Card's work established the mouse for use in the first research personal computers at Xerox PARC and also for the first commercial personal computers (by Xerox):

A key question was whether there were better alternatives to the mouse that we ought to pursue at PARC. I was supposed to help Bill English design an experiment to test this. I always think that there's a problem in just doing A versus B comparisons in devices, which is usually what is done, and what English had done before. So the idea here was to do a model of each possible device, or of each of the devices that we tested. We spent six months putting together a laboratory for evaluating the mouse, assembling equipment and programming a computer to gather motion statistics.

Card began studying human performance with pointing devices. He knew that there are fundamental limits about how quickly people can respond and control devices. For example, there had been studies measuring the rate at which soldiers can aim and fire anti-aircraft guns. Fitts's Law is an empirical law that describes human performance on such eye-hand coordination tasks. Card recognized that fundamental measurements of human performance could be relevant to understanding the mouse. Fitts's Law predicts how quickly a person can move a hand from a starting place to a visible target. The time increases with distance to the target. It also increases when the target is small and hard to hit. At first, Card had a lot of difficulty making sense of the test data on mouse performance:

I was sitting in my office with this data when I first realized that Fitts's Law didn't exactly fit the mouse. When I calculated the slope of the curve, I realized that the *slope* was similar to the *slopes* that were published for Fitts's Law. The difference was that there was a small startup delay to grab the mouse, but once the mouse was in hand, the time to move it to point at something followed the expected curve. Then I realized successively that that meant a person could move the mouse about as fast as they could move their hand. That meant that overall performance was not held back by the mouse. People were going about as fast as they could. The limiting factors for increasing speed were not in the mouse, but rather in the basic eye-hand coordination system of the human. No other pointing device was likely to be better, because the performance limitation came from how fast a human could accurately move the eye and hand. Then I realized that that meant that the mouse was nearly an optimum device. Then I realized that that meant if you introduced the mouse into the market nobody could beat it.

Answering the question about whether the mouse was optimal for pointing performance was only one step toward innovation. When inventions come out of research, there is often a delicate relationship between scientists on a research team and engineers on the development team. In testing the mouse for use in a product, Card faced further challenges:

Having a theoretical model gave me protection against disputes about the empirical findings. I went down and gave a talk at El Segundo to a group of Xerox engineers who were trying to perfect pointing devices.[16] They were very hostile to the mouse because it hung off on a cord and everything at the time was built into a keyboard. The idea of having a part that hung off on the side seemed unmanageable to them. On these early computers the keyboards and displays were built as a single unit. Even the idea of having a separate keyboard was radical.
As I presented the results of the experiment the engineers picked away at the conditions and the control of the experiment and just about everything. Then I

went through why the results were like they were, which came from the theoretical model. They began to see that there was an organized reason explaining why the results came out like they did. It was not likely that there was an artifact in the experiment causing these results. Rather, the theory showed that there was a coherent story of causes. It was logical that the results *should* come out this way. As they understood this, the engineers fell silent.

The mouse was used as the pointing device in the first personal computer that Xerox introduced to the market, which was a version of the Alto. Later, when Xerox developed the Star personal computer, the issue of choosing the right pointing device came up again:

Later there was a big shoot-out over what kind of pointing device Xerox should introduce in the Star. The theory was what really won the day for the mouse. I don't know if there might have been a mouse eventually on the market. We might have ended up with joysticks or trackballs as the standard pointing devices on personal computers. The theory was quite important in establishing the mouse.

In developing the Star, there were additional challenges of making pointing work fast enough with much bigger screens. Because the product organization for the Star product was near the research center, people often got together for lunch. The Star team had redesigned the electronics for the mouse. The issue of performance came up:

In the design of the Star, a question arose about whether the circuits were fast enough for the mouse or whether they would need to use more expensive circuits. Ralph Kimball asked me this question over lunch. I did a quick calculation for the longest possible movement you can make with the mouse, which is the diagonal of the screen. This calculation showed us that the maximum velocity for the mouse was faster than what the circuits could track.

Then we went down to the lab. Because we had good instrumentation for human interfaces we could do an experiment in an hour that spot-checked my calculation. Within an hour or two of hearing the problem over lunch we had an answer to it and told him "You have to make the circuits faster" (at an additional expense). And so they did. There were other systems that came on the market later that had clips for the velocity of the mouse so if you moved the mouse a longer distance the cursor wouldn't go where you wanted it to. That's a fairly subtle problem. We learned to avoid the problem of speed clipping in under two hours of analysis. That's because we had a theory.

Developing the mouse brought many challenges. This story about human performance and the mouse comes near the beginning of the journey for the mouse—the first steps of innovation beyond Engelbart's invention. Was the mouse really better than other pointing devices? How fast did the circuitry have to be? Since then, the mouse has continued to evolve. Optical mice have been developed to increase reliability by eliminating moving parts. Wireless mice have been developed to reduce desktop clutter and eliminate the tangle of cords. Wheels and side buttons have appeared to make certain interactions faster. Many people and companies contributed to the mouse before it became as ubiquitous as it is today.

All Design Is Redesign

Ben Shneiderman is a professor at the University of Maryland, where he founded the Human-Computer Interaction Laboratory. A prolific inventor and a prolific writer, he reflected on how his sense of long roads to get something right was first shaped by observing his parents writing together[17]:

> My parents worked together. They argued and they discussed. They were journalists. My father would dictate and my mother would transcribe in shorthand and retype. And they would just go on arguing.
>
> Seeing them—my father cutting and pasting, where you used glue and paper and scissors and my mother retyping and retyping until it was good enough—made me aware that writing was difficult.
>
> I don't have the usual blocks that I see in other people who struggle with writing. I think there is *the illusion that you have to get it right the first time.* Instead, just push forward and get it down. Then refine, refine, refine. In one project that I did, I had 15 chapters. Each initial chapter took 3 hours. I kept track of my time and the revision time was about 15 hours per chapter. I was quite aware of the level of effort for revisions.

Writing requires massive revisions in order to create a final draft. Repetition of writing and polishing builds confidence and trust in the process. Only the first draft comes quickly, like the first flash of insight for an invention. Most of the time is spent in revisions.

Reflections

Success in innovation is not guaranteed. The tape recorder in Nobutoshi Kihara's story led to the first commercial success of the company that became Sony, but it was not that company's first effort. In the late 1940s, when the company was struggling to become established, they first tried to establish a financial base. This would provide the means for Ibuka to realize his vision of creating a product that was completely new. The company decided to make a rice cooker. The design was extremely simple. Small wooden tubs were fitted with aluminum filaments on the bottom. When rice and water were added to a tub, the water would act as a conductor to complete the circuit. The elements would heat up and the rice would cook. When the rice was cooked, the water would be gone, and the electricity and heating should stop. As Akio Morita explained it in his book *Made in Japan,* the design didn't work reliably:

We bought more than a hundred of these tubs in Chiba and made them into rice cookers. The problem was, in those days you never knew what quality of rice you were getting. With good-quality rice, if you were careful, it came out fine. But if the rice was just a little off, if it was too moist or too dry, we'd end up with a batch that was soggy or falling apart. No matter how many times we tried, it just wouldn't come out right. (p. 13)

When companies present their histories, many of them are inclined to focus on their successes and to hide their missteps. In contrast, Sony's rice cooker has its place in the Sony museum in Tokyo, together with the many Emmy Awards and ground-breaking products like the TR-63 transistor radio, the Trinitron, and the Walkman. Including the rice cooker in the museum addresses the principles that success requires trying something, that you can't know at the beginning exactly what will work, and that it is important to learn and grow from mistakes. This is as important to the soul of innovation as Ibuka's grand vision of creating a company in which engineers are challenged to use their imaginations to create excellent possibilities for a technological future.

A similar story comes from Corning. Corning Incorporated is the glass company most familiar to consumers for products with the brand names Pyrex, Corning Ware, and Corelle. Most companies that have such successful products are tempted to settle back and just produce them, without looking any further. Corning is different. Today, these brands are made by a consumer division that is no longer a part of Corning. Corning has emerged as a major high-technology company leading the world in fiber optics and in cable and photonic products. This transformation of Corning is the result of its commitment to investing in research and innovation to create new business opportunities.[18] Any company that embraces innovation on such a large scale must learn to deal with adversity.

From 1932 to 1935, Corning accepted the challenge of fabricating a 200-inch mirror for Mount Palomar Observatory after General Electric had tried and failed at the project. This was to be a mirror for the largest optical telescope in the world. To manage risk and acquire necessary new knowledge along the way, Corning proposed to work out the process for a large mirror by creating a series of successively larger disks. Starting with the smaller sizes, they created three 60-inch disks, one 60-inch by 80-inch oval disk, a 120-inch disk,[19] and then the 200-inch disk. The project required overcoming many obstacles, including the unplanned crash development of a new glass material to compensate for difficulties in annealing.

On the day that the 200-inch disk was being poured, much of the management team stood in a gallery to watch and celebrate the event. However, 8 hours into the pouring, the weight of the glass and prolonged heat destroyed the cores at the bottom of the mold and ruined the casting.

Like Sony, Corning is more inclined to learn from its failed efforts than to hide them. Corning kept the flawed first disk and placed it in the Corning Glass Center for all visitors to see. Graham and Shuldiner describe it in *Corning and the Craft of Innovation*[20]:

Beyond confirmation of scientific laws and increased know-how, the most significant outcome was the recognition of the value of persistence and the importance of managing failure. That this was a lesson consciously learned by the company as a whole could not be in doubt, for the disk that was not shipped to California, its dusky yellow honeycombed side exposed to view, complete with missing and misshapen sections, was placed first in a square in the center of town and later in the Corning Glass Center for all visitors to see. Unlike GE, which had abandoned even the potentially reusable portions of its project, Corning treated its first flawed 200-inch disk as an emblem of the value of patience in adversity; a reminder of the need to stand firm in the face of failure if ultimate success is to be achieved. (p. 113)

Ultimately, the engineers at Corning were successful. Further struggles were ahead. On the day that the second large disk was poured, the nearby Chemung River overflowed its banks in one of the worst floods in 100 years. Workmen had to scramble to move electrical transformers above the floodwaters to keep power for the annealing cycle available. They succeeded, the casting was finished and the large mirror was created and installed at Mount Palomar, where it is still in use today.

For repeat inventors, seeing their invention in the world makes it all worthwhile. Success rewards determination. Innovation is a test of mastery. Innovators who have taken the long road are seen differently by their colleagues. They are the ones who have made the journey and returned.

II

Fostering Invention

5

Mentoring and Apprenticeship

It all goes from imitation to assimilation to innovation. You move from the imitation stage to the assimilation stage when you take little bits of things from different people and weld them into an identifiable style—creating your own style. Once you've created your own sound and you have a good sense of the history of the music, then you think of where the music hasn't gone and where it can go—and that's innovation.
—Walter Bishop Jr., quoted in *Thinking in Jazz* by Paul F. Berliner

Rite of Passage

When does the education of an inventor really begin? In *The Nightingale's Song*, Robert Timberg tells the stories of five American military leaders prominent in the Iran-Contra affair. Timberg originally planned to begin his account with childhood influences, but along the way he recognized that the most potent influence shaping the thinking of these men was at Annapolis, the U.S. Naval Academy. The Naval Academy created a potent right of passage changing boys into men. Annapolis gave them values and methods that shaped their lives. For invention and research, the analogous rite of passage into professional work life is usually graduate school. It applies to both men and women.

Graduate school is a rite of passage for becoming researchers and inventors. Graduate schools create the next generation of researchers and inventors who are primed to step into positions in the world of science and innovation.

The experience of graduate school draws on a much earlier tradition than undergraduate education, or even high school and grammar school. Education before graduate school is dominated by a program of lectures, exercises, and exams. Such educational practices have a predetermined curriculum intended to serve classes of students essentially in lock step. Education is organized this way in order to efficiently convey the material. The use of lectures saves time in presenting the material to many students at once. Exams are intended to serve as gates that ensure that every one who passes through the system has mastered the knowledge.

In contrast, graduate school is based on the older tradition of mentoring and apprenticeship. Graduate education is about assisting students to take on a professional practice. The curriculum is more tailored. Students acquire the practice by working with multiple mentors, adjusting the emphasis to fit their career objectives. Students discover, sometimes by osmosis, crucial elements of practice that would seldom be encountered in a classroom setting. Graduation requires demonstrated mastery at the level of a practitioner in the field.

Stories

Mentoring

There is often a gap between what can be learned in formal lessons and what needs to be conveyed in total. Paul Berliner described just this effect in his account of learning to play jazz trumpet from a gifted performer[1]:

A trumpet player who once accepted me as student gave me a series of musical exercises to practice. Each time we met, he encouraged me to learn them more thoroughly. When I had finally developed the technical control to repeat them unerringly, he praised my efforts in a manner that seemed to say "That's fine and that's what I have to teach you." The problem was that what I had learned did not sound like jazz to me. When he first sensed my disappointment, he seemed surprised. Then he picked up his instrument and added modestly "Well, of course, you have to throw in a little of this here and there." To my ears, the lifeless exercises I had been practicing were transformed into a vibrant stream of imaginative variations that became progressively more ornate until I could barely recognize their relationship to the original models. This experience awakened me to my responsibility for effecting meaningful exchange between us as teacher and student. (p. 10)

When graduate students begin working with their mentors, they are embarking on a journey with an experienced guide. Apprenticeship amounts to going around the research cycle a few times, asking questions, and getting help at the trickier steps.

Berliner's story suggests that there can be an awakening. In this sense, the preparations are not just about skills. Waking up, however, is often a very gradual process. One can go to graduate school "for the ride" without consciously noticing a shift in attitude. Nonetheless, the experience shapes the student. Because so much of research and invention is about seeing differently, the conditions are set for students to deepen their observations both of their work and themselves.

Squelching versus Stretching the Inquisitive Mind   John Riedl is a prolific inventor and a professor at the University of Minnesota. He has many students at the masters and doctorate levels in computer science and carries out his own research with them. He is appreciated by his students for the care he puts into preparing them for a career in research.[2] Education is a hot topic for Riedl. He believes that education should inspire students, and that mostly, it does not:

There is this great characterization about the American educational system. It says that we teach people for 8 years of grade school to memorize a bunch of stuff and take tests. Then they have 4 years of high school where they learn to memorize a bunch of stuff and take tests. Then there are 4 years of college where they memorize and take tests. And then we put them in graduate school, and give them two more years to learn how to memorize and take tests. Then we say "OK. You are ready to do research now." Not only do they not have a conceptual model for how to do research. We have beaten out of them any capability for open inquisitive curiosity.

To prepare students for research, we start by trying to figure out what their interests and strengths are. If there are a couple of weaknesses that they need to address in order to pursue their topic of interest, I'll suggest that they fill those in.

We have weekly scheduled meetings. The meetings are the nag factor. Students are always thinking that they should have produced something because they are going to have to meet with me and I'll look grouchy if they didn't. But in those meetings it's mostly just them coming in and talking to me about what they worked on, what they have had successes with, or what stumbling blocks that they have bumped into. We spend most of our time talking about those stumbling blocks. If they have strategies to overcome them, then that's great. If they don't, I will see if I can suggest someone to talk to, a paper to read, or a direction to go.

What I do depends very much on the outcomes of those meetings. I have had a wide range of students in terms of their self-directedness. Some students do the best if I leave them alone. They just keep going out and doing cool stuff. There are other students that tend to get lost in blind allies. For those students I set short-term goals. I might say "OK, so for next week you could do this."

When students start out, they don't know how to do research at all. I tell them very directly "Your number one goal in the first year is to write a paper on a research topic that interests you. Pick a project that is one of the dangling leaves on our other projects. Your goal is to write a paper. Not because that is going to be your PhD thesis or where you are going, but because that is going to help you learn how to do research. Then when you are ready to pick your PhD thesis, you will know what it means to pick one."

The first paper is very hard for most students. I have an organized and structured way to teach them how to write a paper. I have an outlining methodology that we follow. It is very detailed down to paragraphs and topic sentences. I don't let them write until we have done the paragraph and topic sentence outline. I review that outline with them for several hours before they put pen to paper. Again, this first time they have no idea what kinds of things you might say where in a research paper. They have read a bunch of papers, but producing one is different from reading one.

Then they write that first draft. We do multiple revisions. Most of the time it is just horrible. Then in the next paper they can usually produce an outline that is a little better. We still spend a few hours working over the outline. Then the next paper I usually find is pretty good.

I am very pragmatic about the value of papers to research. I believe that they have enormous value both for focusing what you are doing within your research project and for understanding which projects you ought to pick. In my experience publishing a paper is very easy if it meets two criteria. It must be something that hasn't been completely done by someone else yet, and it must be interesting.

These criteria are obvious to researchers, but many students don't get it. They want to work on very nice, crisp, narrowly focused-in-a-box problems. Good research problems don't meet these criteria. I can't tell you the number of students who come to me after my course on the fundamentals of operating systems. I love teaching operating systems because it is just such a fun set of ideas. They get all excited. They say "This is just great. I want to do my PhD on an operating systems topic like virtual memory." They don't notice that there's nothing left to do in virtual memory that's really exciting.

I push students very strongly *not* to do a literature search when they are first thinking of ideas. In that stage they should just dream and think and fantasize. They should talk about what *could* be. We will spend a long time, maybe as long as month, just dreaming and fantasizing and imagining. *Then* we do a literature search.

The result is that when we do the literature search, we always find that other researchers have done about *half* of what we have dreamed up. Always. It has never happened any other way. But we have enough energy and excitement so that when we find the state of the art, we can then tune what we did to fill in the holes.

The interactive cycle for a student starts with writing a first research paper is learning about the constructive use of criticism. From the perspective of a professional, the first drafts of a student's paper are usually muddles of poorly understood ideas, technical errors, and claims taken out of context in the intellectual tradition of a field. Accustomed to striving for high grades, students often seek perfection. One of the hardest lessons for a student receiving criticism can be learning how to hear it and use it constructively without dropping into despair and self-doubt. An apprentice's interactions with a mentor represent a trial run for learning to use criticism. As we learn to view and use criticism constructively, we begin to see our critics as valued allies in our search for understanding and clarity.

Mastering material, gaining perspective, and cogent presentation are continuing challenges in a professional career. Shortly after completing her textbook with Carver Mead on *Introduction to VLSI Systems*, Lynn Conway[3] reflected on the process of refining and rewriting the book through several drafts. As she taught classes based on drafts of the book and collected comments from colleagues, she looked for what she called "signals of resistance," places where students misunderstood concepts in the book or colleagues criticized the message. She saw these signals as clues to where further work was needed. Is the conceptual structure wrong or confusing? Are there implicit assumptions that need to be called out for the reader? Are the examples misleading?

Some graduate schools provide a setting in which the lively and semi-public airing of ideas can take place. These settings enable the audience to ask questions, and provide challenges for the presenter to think on his feet and handle criticism. A student or a professor might hold forth with a proposition or a half-baked idea, inviting all present to participate in an exploration of ideas and debate. Since the topics can be wide-ranging and technical, they are bound to touch on issues that new students will not be familiar with.

The exchanges can lead to what seems to be another hard lesson in graduate school. In many cases, new students will remain silent and miss the essence of an argument for fear of seeming ignorant by asking questions. Fully aware of the nuances of this situation, Joshua Lederberg models the asking of questions in colloquia:

At the colloquia I make a point of acting exactly as I did as a graduate student: I pop up with questions. I try to provide an example of not being afraid to appear ignorant or even foolish in asking questions. Sometimes my ignorance is feigned and sometimes it's real—I try to keep people guessing. The main point in the posture of not knowing and asking is an uninhibited search for knowledge.[4]

Notwithstanding his Nobel Prize, Lederberg is always curious and forever a student.

Teaching Graduate Students   Pattie Maes does most of her research at the Media Lab as a professor working with graduate students. Maes's experience as a graduate student shaped how she now teaches her own students[5]:

When I became a graduate student, my professor was very active and still doing research. I was not good at doing research in the beginning at all, but I learned by being involved in the projects that the professor was working on and seeing how he did things. He didn't usually start out with a crisp idea. He went ahead and started coding things and along the way he refined his hypothesis. Where the research ended up depended on what he discovered along the way.

I was strongly influenced by his style and pass it along to my own students. I think mentorship and apprenticeship are really important. One of the reasons I was attracted to the Media Lab was that it was founded by an architect, Nicholas Negroponte. They use the same methodology in architecture. You design something and then you discuss it and change it. They also have the tradition of apprentices and that fits very well with my interests in building computational systems.

When students graduate with their bachelor's degree from MIT, they are not really ready to do research. I was the same way. They do not have a clue about what research is like. Being an apprentice seems to be one of the only ways that you can learn how to do research.

At MIT we have an undergraduate research program. We encourage undergraduates to work in the laboratories. In my group we always have about 15 undergrads that participate in the research. That is one of the only ways to learn about how research really works. Students have to overcome many wrong preconceptions.

Most of the master's degree students, especially the younger ones, don't really know what to work on or how to get started. I think that's because they are used to getting everything neatly presented to them in school. Classes are well thought out and research papers spell out their hypotheses very neatly. Consequently, a lot of our students have a totally wrong idea of what *research* is all about.

When I tell them about a vague idea that I have for a research project, they are often very critical of it because it isn't well defined enough yet. They can't envision how a vague idea can lead into something more interesting and into research papers.

My philosophy is that you just start by building something and you find out where more novel technologies are needed. You do the research and innovation along the way of getting to a particular goal.

Usually I meet with students once a week to discuss what project they are going to work on. We exchange ideas and we decide which ones we are both interested in. They think about that for a week or work on it a little and come back with some variant or some reasons why they think it's not a good thing to work on.

The process is to get started by building. I always encourage the students not to do too much initial reading of relevant literature because it scares them away.

For example, if they start by looking up all the related work, then they often get the feeling that nothing new can be done in that particular research area. I usually like them to develop a system first in a fairly naive and ignorant way. Then we study the literature and I point out the related work to them and we try to figure out what is interesting and different about their work. Then we try to further develop those aspects so that we can come up with some novel technical and scientific contributions.

I think that students have fewer original ideas if they become influenced too early in the process by the predominant researchers in the field. I want them to start with a fresh look at things. I'm not saying that they never have to look at related work, but I want them to do something on their own first. After they've built something, they look at whatever Xerox PARC, or Stanford, or CMU has done in the same area. Then they say "Oh, they see it this way and we see it that way." If they start out by reading papers then they can't easily see things in a unique way anymore. So they can't be as original.

Graduate school consists not only of learning, but also "unlearning." For teaching about the practice of research and invention, professors need to go beyond the kinds of formal instructions characteristic both of scientific papers and what is taught in a classroom. Recognizing this shapes what professors and other mentors need to teach.

As Riedl and Maes characterize it, many students enter a graduate program with an appreciation of those parts of science and engineering where the learning is most formalized and crisp. They have spent years learning how to memorize and take tests. Crisp formalisms are perfect for tests: the rules are stated concisely and they can often be applied unambiguously to solve certain kinds of problems. What they need to "unlearn" is that crisp and formal areas are the best ones for new research. Most opportunities for discovery are at the fringe.

The difficulty is that the *fringe* of invention and discovery is not neat and tidy. Research is the opportunity to *make* them clear. In most cases, the areas where the questions have become neat and tidy are exactly the areas where most of the research is already done, making these areas less suitable for new research investigations. Graduate students actually need new and different skills for exploring the unknown. They need experience both in knowing how to explore and also how to cope with being lost.

The differences between taking tests and advancing the art explain why it is that graduate students almost always start out being lost. They are unsophisticated at picking problems, planning research, or writing up their results. For Riedl, the plan of action is to get students to begin practicing what they need to do as researchers. People get good at what they practice.

Finding a Voice   John Seely Brown has mentored young researchers throughout his career, first as a professor and later as a research manager. He sees the mentoring and apprenticeship relationship as very subtle. It is not just about conveying wisdom and experience to a new generation. That is not enough, because the next generation will almost certainly face problems not faced by the previous ones. Brown believes that a critical part of growing up as a researcher is finding one's own voice[6]:

Young researchers have almost no idea what their real voice is. Most young researchers starting out are fundamentally inchoate. At best, they have some vague intuition that sounds off the wall and often isn't articulated coherently. Part of mentoring is helping people to figure out how to articulate better what their intuitions are, and the second is "finding their voice behind these ideas."

The search for a voice is empowering. And that is something that you don't learn in graduate school, by and large. *Talk* helps a great deal, but there is a sense of not saying too much and a sense of asking the right questions.

There is also a sense of trying to draw out a distinction because we learn to see differently *around distinctions*. So if you launch a distinction, that distinction helps somebody see something about themselves or others that they can't see otherwise. How do you grind new eyeglasses for seeing things differently? You grind new eyeglasses by creating distinctions actually. So you reflect on a situation that they are a part of, and that then starts to shape their sight, and their ability to see, and to see differently.

In some sense, mentoring is a lot like therapy. You help people to discover their own voice by launching distinctions *in situ* that enable them to see themselves and others differently. You create a very safe and encouraging context for that to happen. And in that safety and willingness to suspend your own disbelief, you enter their space as best as you can. Then in the right moment, you see if there is something very simple to say that suddenly turns their perspective.

Brown's focus on helping students and researchers to find their voices reaches the central part of renewal in a field. Education is not just about creating clones of the previous generation. It is about fostering creativity and individuation in the next generation so that it will not only master the lessons already learned, but also so it will be ready for challenges that are ahead.

Coaching a Group   In his role as a professor and director of the Center for Design Research at Stanford, Larry Leifer has shifted his approach to teaching over the years from a more traditional approach in product design toward an open and experimental approach that emphasizes projects. We asked Larry Leifer to speak about mentoring in graduate school, and he told us the following[7]:

Coaching is our label of choice. It touches on all other labels—mentoring, apprenticeship, tutor, advisor, friend, psychiatrist, financial advisor, and so on. The magic of a really good coach is knowing which perspective to apply. As in a lot of human endeavors, it's not just about knowing some facts, but when to use them.

People often start with the assumption that creativity is an individual thing. That idea began to shift for me because of what I was seeing in my own teaching, especially in a course we called megatronics. When we first started these courses, we thought that designers had to be Renaissance men, mastering all of the subjects—electronics, software, mechanics, human interfaces—and doing whatever was needed on a project. It became painfully clear that people couldn't do that. Only one student in a hundred could do that. We switched our approach to creating and coaching *design teams*.

We have somewhat different value systems in our different courses. For master's students and industry-sponsored projects in the product design loft, the value system says that a student must come up with a need that they will then work on. In the team-oriented classes, we have a variant on that. The corporation must come up with a need from the real world and some money. They provide a corporate liaison who communicates what they perceive the problem is that the students should work on. They also provide money to help us run the course.

The students are then invited to challenge everything that the corporate liaison says. Find out what the real need is. Ask the boss of the corporate liaison and check the story out. Also ask marketing and check the story out. The first third of the year is spent on two things: team building and requirement redefinition. And only at the end of three months do we begin to actually try to solve the problem. So there is a big emphasis on problem formulation.

The notion of building prototypes goes beyond teaching. This became clear to me when I got drawn off to look at learning technology. I began to understand that the reason we built these prototypes was to *learn*. That needed to be drawn out for students and even for the faculty because there is an initial preoccupation with the question "Can we build it?" "Does it work?" Only sometimes would people ask "What did we learn from that?" Learning is the primary objective.

The things you see in the lab—paper bicycles and other items—were primary assignments in the team building over the past few years. Typically we do several complete design cycles in a course. The teams form themselves according to our psychometric guidelines. Then they have two weeks to produce a paper bicycle and race it. Sometimes the teams are distributed globally. This last year two teams had members in Tokyo and Sweden. When teams are split geographically, members at both sites have to develop the bike, share the components, and time

Figure 5.1
A paper bicycle developed at Stanford University's Center for Design Research.

the race. The bicycles are often not beautiful, I assure you, but they got pretty good at building them. One of them [figure 5.1] is almost as good as a mountain bike in performance racing around the inner quad.

What we did this year was a rickshaw. That turned out to be a more attainable solution. Everybody finished the race. When it was bicycles, only a third of them finished the race.

One of the rickshaws was very innovative. It took the notion of a rickshaw and abstracted it freely. Unlike most of the designs, which look like familiar rickshaws, this one is more extreme in its shape. It performed very well and the team had one of the most extreme handicaps possible. One guy on the team was 6 foot 6 and weighed about 240 pounds. Also on the team was a woman of about 5 foot 4 and perhaps a hundred pounds, if that. In the race, everybody has to pull the rickshaw. There are three laps—one for each member of the team. They had to make quite exquisite tradeoffs. It was a very nice solution.

The physical space of our teaching laboratory is part of our story. We believe that the physical space should have evidence of our thinking in the world around us both to encourage our reflection about it and to allow others to see what we think and do. I think that the latter may be more important than we've given it credit for in the past.

I think you would recognize this as a reflection of the IDEO[8] space. One of the striking features of this moment is that this design program has done well but it never got famous. We've been small, graduating about 15 people a year. IDEO, the spinoff, got famous. And their approach to design, approach to innovation, approach to cross-functional teams, approach to culture building has become a model, perhaps a standard model. Dave Kelley is now on our faculty here and splits his time between Stanford and IDEO.

A big part of apprenticeship—whether as a member of a jazz band or an engineering team—is learning to work together. Different professions emphasize this to different degrees, but it is an essential part of professional life. When students work on projects, they learn to give presentations and to listen to each other. Project situations mirror later professional situations, where they learn to negotiate through complex patterns of interdependent work involving both competition and collaboration.

Joint projects are an introduction to the skills of working in a social matrix. In a sense, students are joining a guild. Their learning takes place not just with their mentors, but also with each other. When students work together on intense projects and encounter each other in professional settings, they begin to build relationships with their peers. These relationships start a bond that can persist throughout a professional career.

Creating Multi-Disciplinary Studies   Many of the most fertile areas for innovation arise at the intersections of disciplines. John Hennessy, president of Stanford University, has found that cross-training is helpful to prepare people to work on teams across fields, but that students must be thoroughly grounded in at least one academic field. This creates challenges in education, both in the design of graduate programs and in the expected course work for undergraduates[9]:

The majority of our medical students are trained primarily to go into academic medicine or to go into research settings. In that context many of them may find it convenient to get a masters degree in computer science, an engineering field, or in the bioengineering program as part of their education. We will eventually have a set of courses that will help our engineering students who want to go into these areas to be prepared in terms of their knowledge of the biological sciences.

It is important to make a distinction about where studying a second field happens in a person's educational career and to realize that a firm grounding in a primary discipline has to be a key part of the undergraduate program. As you go to graduate school, then broadening and adding different areas makes sense.

A lot of the challenges are about vocabulary. However, we are not going to train everybody to be an expert in all disciplines. You simply can not do it. You can't bring all of the medical doctors to the level of knowledge in physics that you would expect for mechanical engineers. Nor can you bring a mechanical engineer to the level of knowledge about biological systems that you expect of somebody in the medical school.

When I have talked about our goals at Stanford, I have tended to use the term *multi-disciplinary* as opposed to *interdisciplinary*. It's not that the fundamental dictionary based meaning of these words is any different, but I have found that "interdisciplinary studies" is taken with the interpretation that graduates are neither fish nor fowl. What we are really doing in our research projects is bringing together people who think of themselves as neurobiologists or think of themselves as surgeons together with people who think of themselves as computer scientists or mechanical engineers. We bring them together as a team in a multi-disciplinary fashion.

We need a few Renaissance men and women around as the integrators and for inspiration, but we don't expect this for the typical student. That may happen from time to time, but it is certainly not the only model and it is not the dominant model in this situation.

Multi-disciplinary institutions of the sort developed at Stanford and other universities are often crucibles of creativity. They are the locus of crucial new ideas and fields that drive innovation. As a setting for graduate students, they provide two key ingredients. One is that they create a new generation of students whose skills sufficiently cover a combination of disciplines so that they can make progress in the new fields that are emerging. The second ingredient is that they provide examples of the collaborative skills that are needed when people of different fields come together to work on common problems.

Apprenticing

What Is Inside the Door   When Mark Stefik applied to Stanford University as an undergraduate and later for graduate school, he didn't realize how graduate education would shape his life as a researcher[10]:

When I got my acceptance letter as to be an undergraduate at Stanford, I was standing in my laboratory in my basement at home. The letterhead said "Leland Stanford Junior University." I thought "Oh my God. *Junior*! No wonder I got in. I've made a horrible mistake. I've applied to a junior college." Of course, the university was named after Leland Stanford, Junior—the son of the school's founders.

I had applied to Stanford to study pre-med and engineering. As an undergraduate, I kept looking for deeper answers. I gravitated in my studies from biology, to chemistry, and finally graduated in mathematics. I determined that the essence of biology was in chemistry, the essence of chemistry was in physics, and the essence of physics was in math. I had gotten better at extrapolating where this was going. I managed to skip majoring in physics and I'm glad that I didn't see as far as philosophy or I could not have graduated in 4 years. All this time, I earned at least part of my way working at the computer center as consultant on the new IBM system 360 computers.

Before starting graduate school at Stanford, I worked at the Instrumentation Research Laboratory in the genetics department at Stanford medical school. I had always been a quick study with computers. It came very easily for me. Working there, I kept bumping into people who knew about things I had never heard of—actually using basic techniques in signal processing and numerical analysis. I didn't even know how they had learned about these things. I figured that if I was going to be a computer professional, I had to go to graduate school.

I don't think that I fully recognized that people got doctorates to do research. I never conceptualized myself as a researcher, maybe at some level as an inventor. I really admired Edison as a kid. I just wanted to learn everything there was to know about computers. When I got there, there were basically no course requirements. I was different from most of my fellow graduate students in that I took as many courses as I could. I was really hungry to understand things.

Initially I was going to do a thesis in operating systems, but all the systems professors that I knew went away. Vint Cerf went off to DARPA and Forrest Baskett went somewhere too. I was supported by the Heuristic Programming Project which did AI[11] research. I knew people in the genetics department and had a *Scientific American* level background in molecular genetics. Bruce Buchanan, Edward Feigenbaum, and Joshua Lederberg wanted to start some new projects beyond their chemistry research in AI and move into working on genetics. I liked the idea of being on the ground floor for a new project.

I was definitely apprenticing but I wasn't fully aware of it. Ed and Josh were always there setting research directions in a larger context. Ed also got me to write a paper about a DNA segmentation program that I did as a first project at the genetics lab. Bruce had weekly meetings with me as my main advisor, helping me to define a thesis and to write clearly. I was part of a research team with a visiting professor Nancy Martin and another student, Peter Friedland. There were challenging problems in every direction and we were very busy. By the end of grad school I had a whole new set of skills and attitudes. I didn't start graduate school to become a researcher, but it changed me.

Not everything happens in courses or in formal meetings with your advisors. There are a lot of side conversations that you have with your advisors that teach you about the responsibilities and life of being a scientist. At about the time I began graduate school, I remember taking a walk with Joshua Lederberg. I had asked him about the "war on cancer" that the National Institutes of Health was starting to fund He told me that cancer was very complicated and had many different kinds of causes. This was not the kind of problem—like say "going to the moon"—that was likely to yield to a crash program. There wasn't a single place where the research could be focused to get the desired results. He thought that the program might be good politics, but it was badly thought out.

On other occasions, we talked about how science competes for funds with other economic needs. For example, money is always needed for defense, for social programs, and so on. A scientist needs to deeply understand why research deserves a share of the investment. Advances in science and technology are exactly the things that change the way the world works. Potentially, they can make tomorrow better than today. Few other investments can do that. I think that researchers need to understand this in their bones, and be ready to make these kinds of arguments at various points in their careers.

In other conversations, Josh advised me to always work on important problems, rather than trivial ones. At the same time, Ed Feigenbaum often spoke about the relation of science to business. We discussed what the challenges and changes were in various businesses and why expert system technology was important in those cases. In summary, there were a lot of things to learn about a life in science—not just the skills of doing research.

Faculty advisors are more than just teachers. They are also models. Their approach to work conveys attitudes about how to choose problems. What makes a problem important? What makes a question interesting? What is your attitude about sharing half-baked ideas? Do you have any techniques for playing with ideas? The research relationship between a mentor and an apprentice can evolve into a meaningful relationship between colleagues that can last for years.

In 1977, Americans won many of the Nobel Prizes. The strength of American science was well recognized. Donald Kennedy quoted his Swedish colleague Suny Bergstrom, who pondered why this had happened.[12] Bergstrom attributed it to the "democracy of American science," by which he meant the fellowship of the bench. As Kennedy said, Bergstrom was referring to "the system of apprenticeship that is built upon the coexistence of research with research training."

Apprentices learn what the life of research is about. Students have to know deeply why research and innovation matters—both to society at large and to particular businesses and funding organizations. In the course of a career, scientists will be called on to defend or explain why the work is important. What kinds of arguments have been made before? Which of these have proved right? Your reputation as a scientist will depend on good judgments about what kinds of results research will lead to. You need to know what is important, what is trivial, and when a crash program can work. Dealing with these issues is part of the responsibility in a life of science and technology.

Apprenticing in Graduate School   The role of thesis advisor can be a "lifetime appointment," the start of an enduring professional relationship. When Stu Card was a graduate student at Carnegie Mellon University, his thesis advisor was Allen Newell. Newell and Card continued to have weekly meetings by phone for many years, discussing their mutual research interests in computers and psychology. Card[13]:

I took the Strong Interest Test to see what area I should study. The results showed people and natural science—which suggested psychology. So I actually took a psychology course. I did not realize how primitive it was as a science at the time. But the AI stuff was a way to work on the mind and behavior that was very technical. Carnegie Mellon had a Systems of Communications and Science program, which was supposed to provide experience from a lot of these areas—everything from small groups to operations research to various kinds of math modeling across all these areas.

One day I heard a talk by Herb Simon and afterwards a guy I worked for set up a session for a bunch of us to talk with him. I was really enamored. I applied for the program to work with Newell and Simon. They ended up putting me in the psychology department.

Sometime in my first year I discovered that they were treating me as if I was a psychology student. I worked for Herb Clark, as a research assistant. I marched up to Newell's office and said "Hey, I got accepted for the Systems and Sciences Program. When does that start?" Newell said "Sorry. This year they ended it." They had accepted me to a program that ended the year that I arrived. I said "That's not fair. I don't want to be a psychologist!" So he said that there were a couple of choices. I could invoke a grandfather clause to do the old program, or I could create my own program.

Newell and I created a program combining psychology and computer science. We got rid of half of the psychology qualifiers, developmental psychology, social psychology and stuff like that and substituted computer science qualifiers such as programming languages. I did that to the utter amazement of the psychology faculty. They couldn't believe that a student would come in and rearrange all of their qualifiers and accreditation structure. But my secret was who I had on my thesis committee: Allen Newell, Herb Simon, and Lee Gregg, who was the head of the psychology department at the time. So there was basically nobody who could stand against this. I just had to convince Newell. He brought the other two guys along.

Like me, Allen Newell majored in physics. That is why I chose him instead of Herb Simon. Simon was a political scientist. He really thought about this stuff differently than Newell and thought about it differently than me a little bit. I have great admiration for Simon's view, but I had this great impedance match with Newell, as you would say in physics, because we had so much common background.

At an earlier part of his life he had been an organizational scientist. So he had this combination of physics plus he knew about organizations. He was a masterful organization man. He would always invent things on a grand scale. He would think outside of the regular structures. I got some training watching him do that.

I would go and ask him "What should you do here? What would you do here? How would you think of it?" And he would analyze it for me.

Newell was very important to me as a mentor because he had this background and he was also willing to take me on. After I graduated, he consulted with my research group at PARC for 20 years. He could have consulted with anybody he wanted to. In the last years of his life when he got cancer and he was too sick to come out here, every Thursday night we talked for two hours on the phone. He had a slot for me. He's an extremely busy guy, but he was willing to give me that much time.

The way you learn science is partly by apprenticeship. If you look at who the really great scientists are, they are usually the students of really great scientists. That is not an absolute rule, but you can do these family hierarchies that are amazingly elaborate and you know that great scientists have other great scientists as students.

Newell had a bunch of heuristics for choosing problems. He had a set of scientific values which I more or less share partly from his influence.

He had a whole catalogue of aphorisms. I will just go through a couple otherwise you will be here for awhile. A lot of these are pretty specific.

A lot of people like to write "framework papers." As you become older and a senior scientist in your career, you get opportunities to give all of these talks and basically become a Grand Poo-bah. He would limit his philosophizing to ceremonial occasions. "Framework talks don't count as getting science done."

Another aphorism is "All science is *technique*; the rest of it is just commentary." This is about what counts as real progress in science. Science is about how you can do something or how you can predict something. When you invent a new scientific technique, other scientists can then use it.

For Newell, science was not a job or an occupation. It was a *calling*.

Absorbing Culture at Graduate School   All professors prepare their graduate students for a professional life, but they don't all do it in the same way. When Daniel Bobrow was a graduate student, his advisor was a new assistant professor at MIT named Marvin Minsky.[14] If Minsky were a swimming instructor, his approach would not be to introduce the swim student slowly to water and teach the basic strokes. It would be to throw the baby in. He provided the pool. Bobrow[15]:

I never thought of Marvin Minsky exactly as a mentor. If anything he built a creative environment for learning rather than providing direct mentoring. Marvin had just finished as a Junior Fellow at Harvard.[16] I think that affected him, in the sense that it was an environment, not a process. And the environment was not just at school, but also his home. Minsky bought a huge house over in Brookline where he invited graduate students to socialize with colleagues like John McCarthy, Warren McCullough, Ed Fredkin, and Jerry Lettvin. More than anything else, Marvin recommended books to read. He provided an environment where there were people around that you could discuss ideas with.

Minsky's first child, Margaret, was very young. They had this huge living room and arranged boxes of books around the carpet in the center to form a giant playpen. All of the other seating arrangements were around the outside. The living room must have been 20 by 30 feet, so it had lots of seating as well as two grand pianos, with the center being this area in which Margaret could crawl. They had these beautiful oak beams up in the ceiling. Marvin sank two huge bolts in the beam to support a trapeze that people swung on over the middle of the room. The conversation room was a play room. We would go over there and just hang out and talk.

What would happen was that people would describe what they were doing. I remember Ray Solomonoff talking about trying to understand what induction meant, and the program he was developing to do learning by induction. I learned so much from just hearing people think out loud, and being able to question them. It was a local school of the Socrates sort, with a couch instead of a log. When Seymour Papert became part of the community, we started discussing Piaget with him[17] and whether Piaget's stages of child development were the right way to think about artificial intelligence Meanwhile, Margaret was playing on the floor. She was a very quiet child and didn't say anything. I don't think that she said a word until she was two or two and a half. Marvin told me that her first sentence was "The steam is coming up through the radiator to keep us all warm." She had been listening, absorbing the language, and she finally had something that she wanted to say.

We didn't have meetings. We had parties. There was food and discussion. There were many rooms in this house and you could always go off and find a place to talk.

I know that some people meet with their graduate advisors to talk about their thesis. I never had a meeting with Marvin about my thesis. We did not actually talk about my thesis until I wrote a draft. We talked about everything else.

When I gave him my thesis, which was about an 80- or 90-page document at the time, he kept it for a while. When I finally got it back every page was covered with red marks. He insisted that I do various kinds of background checks and scholarly this and that and that I explain things more clearly. In some sense it was demolishing. I asked "Why are you making me do this?" He said "Because you can." He was telling me that I had the ability to implement his suggestions and that is what he wanted me to do. Several months *later* I felt very good about it. The thesis was much better. After I re-wrote the whole thing it was about 50 percent longer and was well structured.

I talked to him much later and he was very proud of the fact that he had never written a paper with one of his students. His students had all written their own papers. Minsky thought that lots of professors just tried to get their research done through their students and he wanted to get his students to do their own research.

Probably the only person at MIT that I talked to in any depth about my thesis with was Seymour Papert. I met him one night about 10 p.m. and we talked until 2 or 3 in the morning. We discussed what I had to do in order to make it clear. I needed to explain what was going on in transforming the language of word problems to algebra and why my methods seemed to work. Another early influence on my thesis was Herb Simon. I talked with him on several occasions. There was this sense that the entire AI world was our community.

When Marvin casually said "Oh, come with me to lunch," it turned out that he was having lunch at the faculty club with Norbert Wiener. Wiener was very interested in talking about a science fiction story he had just written. What was amazing was to see was how delicate his ego was. Here was this man I knew as a giant and he would say "What did you think of this story, and what about this idea?" It was helpful to see his human side, someone who was clearly brilliant beyond anything I could ever imagine, having an ego as delicate as mine.

John McCarthy was another giant. He was busy trying to figure out Lisp.[18] He ran a seminar every week describing what he was doing. People would just come in and make suggestions. So it was an open environment. I kept encountering people who were wrestling with very big ideas.

Faculty advisors are also gatekeepers. They introduce students into a world of influential researchers and also into the world of research funding. Part of the learning in graduate school comes from being at the place where ideas are being formulated, where plans are being made, and where new things are starting to happen. When the faculty lead the creation of something new, their students get a sense of what it means to be a revolutionary in charting out new and important directions for research.

Working for an Inventor   When Thomas Edison opened his laboratory in Menlo Park, New Jersey, many of the people who came to work for him started out working for free. They began as apprentices to learn about invention until they became useful. Some of them eventually went on to be inventors at the other research laboratories that were being created at the time. Even today, graduate school is not the only path for inventors. Chuck Thacker is a case in point[19]:

I started my undergraduate education at Caltech majoring in physics. I was going to design particle accelerators, which I thought was a combination of great science and great engineering. I worked my way through school from the time I was married at 21. I started at a couple of companies in Los Angeles. Then I moved to the Bay Area to go to Cal[20] for the last two years. When I graduated from Berkeley, I actually planned to go back to graduate school, but I wanted to work for a year. I worked for Jack Hawley. Hawley was the only inventor in the Berkeley phone book.

I learned an awful lot from Hawley about mechanical design, because that was what he was good at. He did not understand electronics at all. Did I learn anything about methodology? No, I don't think so.

Hawley was a friend for many years, but I would not describe him as a mentor. He decided to be an individual inventor and had a little store front shop in Albany right next to Berkeley. I went to him and said "OK, Jack. I will make you a deal. I will work with you for a year. You can pay bare minimum wages. I will design your electronics for you (which he was not good at and I was) and you can teach me how to run all of the tools in your machine shop." He agreed and that was a fine deal. I became a reasonably good machinist. You could turn me loose in the shop and I could probably make almost anything.

I will tell you about the smallest government contract. Hawley had a friend who was a fairly high bureaucrat in the Army Corps of Engineers. This guy had a crazy idea that he wanted to try out and he got Hawley to do it. The war in Vietnam was revving up at that time, and one of the difficulties with conducting wars in that part of the world is that there weren't any airstrips. The Army had these great C5s that could carry cargo and land on really awful landing fields. But there was a problem if the ground was very wet and saturated because the C5 would sink in the ground. This guy had the idea that you could parachute down a little robot. The robot would crawl around and poke a probe into the ground to find out how soft it was. The results would be radioed back up. Just as a little frill he machined the mold that did this so that as the vehicle moved forward in its wake it would print Hawley on the ground.

If the ground was too wet for the C5 to land you would lose the robot. We built one together during that year. It was a tank with treads powered by two large truck batteries. The probe was a hydraulic cylinder that made it look like it had a smokestack. There were a lot of little things that you had to get right to make it work. Hawley was very good at solving little problems. For instance, how do you make the treads for it? You can't just go to somebody who makes tanks and say "Please make me a miniature tank tread." He got a stainless steel conveyer chain, like you see for carrying dishes in cafeterias. He developed a process to take uncured rubber and rough-up the service of the tread then pressure and heat the rubber to glue it to the tread.

This guy was absolutely unbelievable. He wore a 55-carat amethyst, carried a stick, and wore a bowler hat everywhere. He ran for mayor of Berkeley one time. Berkeley has elected some pretty weird mayors. They didn't elect Hawley. He was enormously flamboyant and would do anything to get his name in the paper.

Hawley had a friend who was a distributor of some sort. He had called on the computing group at Berkeley and happened to mention to me that they were looking for an engineer. I thought it would be nice to have a real job when I went to graduate school instead of working for this crazy inventor. I went and talked with them. They were building a time-sharing system[21] for the SDS 940. I joined this project as a hardware engineer. That was 1967, and many of us have been working together since then.

Although Chuck Thacker did not go to graduate school and did not consider his first inventor employer a mentor, his year of working in Jack Hawley's shop had many elements of apprenticeship. For example, he worked at a minimum wage in exchange for being taught about the machine shop. Along the way he learned a lot about mechanical engineering and perhaps about the attitude of an inventor who knows how to choose and attack small problems.

Interning in Research Labs   The mentoring and apprenticeship roles are not limited to graduate school. When people change fields or expand their interests, they often find other settings that support lifelong learning.

Most research laboratories have programs providing research experience for interns. Some labs offer internships during the summers, and others have intern programs that run all year long. These programs provide a mentoring environment similar to graduate school, except that the intern works on a project with a team of researchers and learns more about teamwork. Ron Kaplan leads the Natural Language Theory and Technology area at PARC, which takes on interns for multiple-year engagements[22]:

We usually have about two interns at a time and keep them for a couple of years. Some of them at times have been assigned to specific things, like grammar writing. For example, two that we have now have been here for a couple of years. Tracy King had these assistants and actually one of them is still working with Tracy. They basically do two things. They keep us a little more connected to what is going on at Stanford, things we're not leading or monitoring all the time. They raise interesting questions. They pick out issues that might not be central for us, but yet are interesting. They are a very important part of our intellectual life.

We have tended not to have summer interns. We have students. We don't like the summer program for the most part because there's a pretty high entry cost to understand all these mathematical and other things that are going on for our projects. In the summer program, you barely get someone up to speed where they can be useful and time is up. About 10 or 15 years ago I decided that this was just not worthwhile. So we tend to try to pick people who are local—students at Stanford or Berkeley—and work with them. We get them maybe in the second year of grad school and keep them for their third year. Then they go off to focus on their thesis. Maybe they'll do their thesis with us as advisors. That's been very important.

Many research centers have intern programs, often with positions both for graduate students and undergraduate students. In the course of their education, some graduate students use these opportunities to gain experience working with professionals in their field and to become acquainted with different companies and research environments.

Finding the Right Questions   David Goldberg reflects back on when he was a student, and how things are different in research[23]:

In school, you take classes and do homework problems. It's not really like working on research. Solving a research problem is different from doing an exercise.

In research, you have to ask the questions, not just have an answer. Also, on a textbook problem, you pretty much know both that there is an answer and that it can be found using the methods preceding it in a book. In research, there may not be an answer or it may require you to find a method that you don't know about or maybe nobody has invented before.

David's observation nails one of the hardest skills in research: Knowing how to formulate the right questions. Joshua Lederberg: "The discovery of an important question is far more important than that we get the answers to what we think the questions are."

Apprenticing with Colleagues   Apprenticing doesn't necessarily stop at the end of graduate school. In a research laboratory, researchers often work with people in other disciplines and sometimes mentor with people in a field that they are moving into. Diana Smetters is a researcher in the security group at PARC. As she made a transition from neurobiology into cryptography, she was mentored by colleagues[24]:

When I have a hard puzzle to work on, especially if I need anything involving crypto, I go to someone like Matt Franklin and I say "OK, I have this problem." Sometimes I say "I have this problem and I have this solution." He used to always say something like "Yeah. That was solved in 1982 and in this paper."

But now I've moved my way up in mathematical sophistication. These days Matt may say "Oh, good. That was solved in a paper that will be published in two months." I finally got one where he said "I've never seen anything about that. Maybe that would work." That was a good day. I seem to be moving toward the bleeding edge.

Matt gave me a pointer to a paper that he thought might be applicable although it didn't solve the problem. I thought I was running into too many problems that were over my head. So I took that paper on Friday to Ralph Merkle and Jessica Staddon and said "Here is a problem that I have. Would you guys be interested in working with me on this problem?" So I'm now working with them on that piece.

Smetters's experiences in learning from her colleagues are not usual. In a multi-disciplinary research environment, researchers are always apprenticing at some level with their colleagues as part of learning to work together. Sometimes people change fields as they find that they enjoy other parts of the work.

Mentoring High School Students

In the United States, there has been a long-term trend where fewer students are making their way into science and engineering. Especially in engineering, enrollment has been dropping substantially:

A long-term trend has been for fewer students [in the United States] to enroll in engineering. . . . Undergraduate engineering enrollment declined by more than 20 percent, from 441,000 students in 1983 (the peak year) to 361,000 students in 1999. Graduate engineering enrollment peaked in 1993 and continues to decline. (*Science and Engineering Indicators—2002*, pp. 2–3, 2–17)

Over long periods of time, different countries in the world have experienced either a rise or a fall in scientific education. The present decline in interest by American students in science and engineering is not matched by a comparable pattern overseas. To put it simply, engineering in Asia is gearing up as the United States gears down. Asian countries produced approximately 450,000 (about six times as many) engineering degrees in 1999 as the United States.[25] This is a dramatic change from the previous decade. Developing Asian countries have greatly increased their production of engineers and scientists at all levels including the doctorate level. Similarly, European countries during this period have increased their production of science and engineering degrees, and almost doubled their production of doctoral degrees.

The trend in science and engineering degrees is matched by trends in test scores before college. In the United States, assessments of mathematics and science performance have been a concern for many years. Only 17 percent of twelfth-grade students in the United States scored at the "proficient" level on the NAEP mathematics assessment in 2000. Perhaps the most telling indicator is the comparison with students overseas:

Internationally, U.S. student relative performance becomes increasingly weaker at higher grade levels. Even the most advanced students at the end of secondary school performed poorly compared with students in other countries taking similar advanced mathematics and science courses. . . . On the general science knowledge assessment, U.S. students scored 20 points below the 21-country international average, comparable to the performance of 7 other nations but below the performance of 11 nations participating in the assessment. Only 2 of the 21 countries, Cyprus and South Africa, performed at a significantly lower level than the United States. (ibid., pp. 1–15, 1–18)

For a variety of reasons, American students deciding what to do with their lives have not been inspired toward science and engineering. This kind of shift is not limited to the United States and has occurred in other countries before. It may be related to larger social and economic cycles. A similar pattern developed in England during the 1970s and the 1980s. It arises when high school students cannot imagine working in those fields and believe that there is no future for them in those fields. The following stories reflect on this problem and also on the results of extending the idea of mentoring to younger students.

Innovation and Education   Research and engineering fields constantly need to be replenished. One effect of the boom-bust cycle in technology is that science and engineering fields look like unstable career options. Fewer students are interested in careers in science and engineering. As the president of Stanford University and as an engineer, John Hennessy is concerned about the trends in engineering:

We are not producing enough young people in the United States with an interest in science and engineering. . . . I think that there are a variety of things underneath this trend including role models, problems in the K–12 schools, and our inability to articulate why these are exciting careers. We have television shows about doctors in the emergency room. We have shows about lawyers, and we have shows about policemen. What shows do we have about innovators? *Home Improvement* is the closest thing that we have to an engineering show, and it's not particularly appealing.

We need to convey more of the excitement and fascination of science and technology. The opportunity for innovation is absolutely remarkable and the opportunity to do really wonderful things is out there, whether you do it in a research setting or you do it in a company. There is a thrill and adventure in creating something new, seeing that through to completion and then seeing your work change the world.

There are also problems in K–12 education that need to be fixed. We need to recognize that math and science teachers are in short supply, and we should pay them more money. If you look at the history of this country and you look at the waves of immigration that have come along, the one thing that you can really see is that education is the path that has helped them achieve the American dream. I think that is something that we have to continue to hammer home.

In one sense, the shortage of students entering engineering and science is at the front end of the innovation ecology because it is about people entering the system. However, there are no front ends or back ends in an ecology. The parts of the ecology all affect each other. When students see so many products coming from overseas, they wonder if they can get good jobs in the United States. International companies like Flextronics specialize in setting up advanced manufacturing facilities in the developing world. Flextronics has plants in Mexico, Hungary, and China. Similarly, many companies in India do technical work at much lower prices than in the United States. Because the international economic and salary gaps are so big, the pressures won't go away overnight. Globalization will increase and the exporting of jobs will continue for several decades.

Inspiring Younger Students Hennessy's analysis is that younger students are not being inspired to enter the field. But does it have to be so?

One of the bright lights in the United States inspiring students toward science and engineering is FIRST (For Inspiration and Recognition of Science and Technology). FIRST sponsors a robotics competition, which has involved more than 20,000 high school students since it was founded in 1989. Dean Kamen, an American inventor known for his medical inventions and for the Segway personal transporter, founded FIRST to encourage students to excel in math and science.

Mark Yim leads the modular robotics group at PARC. For several years, he has volunteered his time to work with students and faculty at Gunn High School in Palo Alto, helping them in their projects for the national robotics competition. Yim found that a project focus opens opportunities for students to discover their interests and abilities. Yim's interactions with students in the robotics competition may not reach as many students as John Hennessy has in mind, but it shows how collaborative experiences on an engineering project at just the right time can have a powerful influence:

There are two key elements in whatever I do: building interesting things and mentoring. Especially for kids that are just about to enter college, you can have a big impact on their lives and what they do. This is a decision point in their lives for what careers they will pursue. Many of the Gunn robotics students say that the robotics experience was life changing for them. Sometimes kids are not necessarily that good at studying, but they get really excited about robotics. They would work on the projects for hours. We had to set limits so that they would go home by 11 o'clock at night, or they would have worked on the project 24 hours a day for the six weeks of the program. There was one kid who snuck out of his house in the middle of the night long past his curfew to go to school to work on the robot. You worry about kids sneaking out to do drugs or something, but here was a kid sneaking out to go to school. It was unheard of.

As the program developed, the parents realized that their kids were spending all of their time there and not eating right. The kids would just grab something to eat so that they could keep working. So the parents organized themselves to bring dinner to the school for the kids on the teams. This way they at least knew that their kids were eating right. At dinner time parents would bring big vats of things like spaghetti, which was really nice.

Several of the students now say they are going to go into engineering, which they had not thought about before. There were some kids that already knew they were going into engineering, but now they know specifically which field they will pursue. For some of them the robotics experience really sparked an interest that was not there before.

Part of the excitement is that it is cool to build a big robot and see it competing with other robots, but there is also a social aspect. The kids bond when they work on a very intense project. It used to be that kids working on the robot competitions were considered weird and geeks. That's changed. Now you have the jocks on the robotics team and also popular kids joining in, all working together.

They have to learn to work together, to solve problems, and to manage each other. They form groups and have leaders and subgroups. They learn how to present ideas and how get buy in and how to do it all together. They learn in ways that they might not experience even in college.

The students used to do more of the fund raising. Now the parents are mostly responsible for fundraising. It was really interesting when the students would go out to solicit money from companies. They acted as if they were a startup in Silicon Valley. They would go out and say "We need to raise forty thousand dollars." The amazing thing was that they would raise more money then the entire math department at school.

One thing that we get back from high school kids and college interns when they work on projects at PARC is fresh points of view. Their spirit is often rejuvenating for researchers. Sometimes they think of things we have thought of before, but almost always, they come up with new ideas. High school kids really know the current consumer technology a lot better then people our age. They all have cell phones and use all of the features. They often know how to use computers better than their teachers. They all do instant messaging. Kids are the users, consumers and researchers of the future.

Yim's experience with high school students shows that technology is not inherently unattractive. FIRST has had some major successes. In 2003 there were more than 800 teams participating nationwide and internationally. In competitions across the country, the contest highlights education in science and math and introduces mentoring and teamwork into the high school curriculum. The national championship event in 2002 drew more than 300 teams and more than 25,000 students to Disney World's Epcot Center in Orlando. The program attracts both boys and girls, some from disadvantaged communities.

FIRST is perhaps the biggest program of its kind reaching high school students. Successful as it is, some additional numbers may put it into context. In the United States, approximately 61,000 bachelor's degrees were awarded in engineering, 105,000 in the natural sciences, 185,000 in the social and behavioral sciences, and 40,000 in computer science. This suggests that FIRST is reaching only a fraction of the students who have already chosen to go into engineering and science. From a disciplinary perspective, FIRST emphasizes computing and mechanical engineering.

Reflections

A central theme in the education of inventors and researchers is that the appropriate education for "doing research or invention" requires skills and experiences that are different from what happens in the usual classroom or what is dominant in formal scientific and pedagogical writing. Formal education and formal scientific writing usually revolve around the explication of results and theories—that is, the formal knowledge products of scientific and engineering investigations.

The word "theory" usually connotes a formal way of thinking logically or mathematically. In this formal sense, theory takes its place in a knowledge-generating process called *the scientific method*. The scientific method includes hypothesis formation, experiment planning, execution, and data analysis. In this scheme, theory is used to make predictions. Theory is created by a process that includes hypothesis formation and testing.

Unfortunately, this notion of theory and the working methods of science and invention leaves out imagination. This makes it both boring and misleading. Peter Medawar was a well-known medical researcher, a noted author, and a philosopher of science. He won the 1960 Nobel Prize in physiology and medicine with Sir Macfarlane Burnet for their joint work on the theory of acquired immunological tolerance. Medawar was a champion for explaining science clearly and conveying the excitement of scientific discovery. No stranger to the scientific literature, Medawar found himself disturbed by scientific papers as representations of scientific thought. He wrote an article titled "Is the Scientific Paper Fraudulent? Yes; It Misrepresents Scientific Thought."[26]

Medawar describes the structure of a scientific paper as including the following sections: introduction, previous work, related work, methods, results, and discussion. The introduction describes the general field. The section on previous work lays out the related work of others; in it, "you concede, more or less graciously, that others have dimly groped toward the fundamental truths that you are now about to expound." The section

Figure 5.2
Two paper rickshaws developed in a course on team-based product innovation
in Stanford's Department of Mechanical Engineering.

on methods describes carefully what the experiment did. The results
section lays out the factual data that were gathered. The discussion
section adopts "the ludicrous pretense of asking yourself if the information you have collected actually means anything." This is where the
scientific evidence is appraised.

In Medawar's view, the standard scientific paper promotes an error of
understanding how science is done because it confuses proving with
discovering. The paper's structure makes it appear that the doing of
science is the careful laying out of facts. Once the facts are in, the conclusions follow. This makes it seem that science is all about deduction.
Unfortunately, this formal structure leaves out the creative part of
discovery and invention. The structure of a scientific paper is only about
proof, promoting the systematic marshaling of evidence. In this abbre-
viated story, once a scientist has somehow gotten some results, the paper
lays out the conclusions logically: "I took the measurements carefully. If
XYZ was true, the results should have shown ABC. But they didn't.

Figure 5.3
A teaching lab at Stanford used by the Center for Design Research. Hanging from the ceiling are paper bicycles and rickshaws developed by students. Standing at the clover-shaped table are Larry Leifer (left) and Mark Cutkosky (right).

What remains is that TUV is the only hypothesis consistent with the data."

This structure for a scientific paper is similar to the structure of an advanced textbook in mathematics. In a typical textbook, the first chapter is a rigorous presentation of crisp definitions, axioms, and theorems. Such presentations have evolved as a style of pedagogy intended to quickly and carefully introduce the logical structure of a complex body of mathematical knowledge. Interestingly, the shortcomings of this way of explaining discovery and invention have also been bothersome to many notable mathematicians, and this has resulted in several books about how mathematical intuition and discovery actually work.[27]

Similarly, most scientific papers fail to describe how scientific discovery happens. What is reported in a regular scientific paper is mainly about how competing hypotheses have been tested. The testing of hypotheses is largely about proof. Proof is a logical and rigorous process based on deductive arguments. What is missing is an account of where hypotheses and inventions come from. Medawar put it this way: "Hypotheses arise by guesswork. That is to put it in its crudest form. I should say rather that they arise by inspiration; but in any event they arise by processes that form part of the subject matter of psychology and certainly not of logic."

For Peter Medawar and others, *discovery is based on imagination and proof or hypothesis testing is based on logic.*

In graduate school, future researchers and inventors begin to sort out these issues. Rote learning and standards of proof—the formal parts of science—are important in creating a field that securely builds on its experience. At the same time, however, imagination and curiosity and the skills of working on multi-disciplinary study teams are important for fostering the creativity that drives the next round of innovation. By working with mentors, students acquire not just knowledge and methods but also values and attitudes. By working on teams, especially teams that mix scientific and engineering perspectives, they learn the elements of radical research. In this way, students gain the skills they will need as professionals.

6

The Prepared Mind and the Aha! Moment

In the field of observation, chance favors the prepared mind.
—Louis Pasteur

Science has many stories of chance observations that have led to important discoveries. One of the most celebrated of these discoveries was made by Alexander Fleming. In 1928 Fleming noticed that a mold culture growing on a Petri dish in his laboratory secreted a juice that killed bacteria. Around every culture of the mold, all of the bacteria had died. The mold was *Penicillium*, and eventually this discovery led to the "wonder drug" penicillin. Fleming's prepared mind recognized an opportunity in the patterns of dead bacteria on the Petri dish. Absent a prepared mind, Fleming could have dismissed his discovery and thrown away the sample without noticing its potential as an antibiotic. This story is a classic illustration of Pasteur's idea that in the field of observations chance favors the prepared mind.

Preparation is essential. It brings the necessary ideas to mind at the same time, so that they can be combined in a novel way. When the combination happens, it leads to a sequence of insights and to the "Aha!" experience of discovery.

Inventors often describe their "Aha!" experiences with a sense of wonder, saying things like "It just hit me" and "I know the answer!" Such accounts give rise to the misunderstanding that inspiration comes simply and inexplicably from genius without previous work. In *Thinking in Jazz*, Paul Berliner describes a similar misunderstanding about improvisation for jazz musicians:

The popular conception of improvisation as "performance without previous preparation" is fundamentally misleading. There is, in fact, a lifetime of preparation and knowledge behind every idea that an improviser performs.

Stories

Preparing the Mind

The journey of a young jazz musician takes place in a community where musicians listen and practice in different settings. Gathering these experiences is an essential part of preparing the mind for improvisation. Berliner:

Amid the jazz community's kaleidoscopic array of information, students glimpse varied elements as they appear and reappear in different settings and are interpreted by the performers whom the students encounter. Learners synthesize disparate facts in an effort to understand the larger tradition. Gary Bartz "basically learned one thing" from each of the musicians who assisted him—"saxophone technique" from one, "dynamics and articulation" from another, "chords" from a third. Similarly, an aspiring pianist learned the general principles of jazz theory from Barry Harris, discovered "how to achieve the "independence of both hands and how to create effective left hand bass lines" under Jaki Byard's tutelage, and expanded his repertory with someone else. (p. 52)

Alan Kay once said that point of view is worth 40 IQ points.[1] One way to look at repeat inventors studying across fields and jazz musicians studying across styles is that they are preparing their minds for invention by accumulating points of view.

Characteristics of Innovative Students   John Riedl is a professor at the University of Minnesota. He characterizes himself as a serial inventor, working on inventions in various subfields of computer science and electrical engineering, including signal processing, software engineering, and collaborative filtering. Many of Riedl's joint research projects grow out of his role directing students in their course projects and doctoral dissertations. Riedl is reflective about the qualities that dispose students to being innovative in their projects[2]:

Some students make themselves *available for creativity*. One of the things a student must have is a willingness to invest in developing a set of skills, not knowing which skills will be useful. I often find that the students who have been most creative are ironically the students who spend the most time learning how to use tools. The students who have "wasted" dozens of hours configuring EMACS[3] to be just perfect for them often turn out to be students who, when they find something, can really go places with it. On the other hand you do have to coach them. (laughs) These are the same students that need help setting goals for the next week. You have to do this because they have divergent minds that will take them away.

The other characteristic I notice is students who prepare themselves by absorbing very broadly. These are people who have lots of life experiences, who read lots of science fiction, and who read lots of non-science fiction such as biographies of scientists. I keep a shelf full of biographies of scientists to hand out to students who might be interested in what it was like for Turing[4] or what it was like for Ramanujan.[5]

I think that it fertilizes your brain so that when the creative thing comes, your imagination is used to being wild and nutso. Obviously other things are important too. You have to have discipline; you have to have skills; and you have to have brain power. But I find that these other activities such as a wide variety of life experiences and reading broadly help students to get prepared.

Deliberate Cross-Disciplinary Study   Invention involves *combining* ideas—especially points of view from different fields. Different points of view bring new ways of thinking. New points of view can lead to shortcuts such as solving a problem in one field by making an analogy to a problem solved in a different field.

Pattie Maes is acutely aware of the importance of finding new ways of looking at things. She takes inspiration from fields that study complex systems. When we spoke to her, she was on a sabbatical from her teaching responsibilities, and was using the time to take courses in other fields[6]:

I often put two very different things together such as a technique from one domain used with a totally different type of problem. For example, I may take an interesting idea from biology and marry it with a problem in artificial intelligence.

I started paying attention to biology about 15 years ago and over time have been looking at other fields. I often don't attend the conferences in my own field, except if I have to present a paper or give a talk. I prefer to attend talks and conferences in other areas. When I was working at the artificial intelligence laboratory with Rodney Brooks, all of us—Brooks included—would study other fields, specifically biology and ethology,[7] the work of Lorenz and Tinbergen. Those ethologists had very sophisticated models about how animals decide what to do and how they learn. They take extensive observations and then make models to explain what they see. Their models were unknown to the artificial intelligence community and rarely exploited in computational systems. That was a rich area to explore and to get inspiration for artificial intelligence and for building computational systems.

I always read books in other domains and learn about them because I will get fresh ideas and see the common themes more clearly. This works better than just reading books in my own field where I will learn a lot but I won't think of totally new things to do. For example, I am currently on leave from MIT and I am taking classes in genomics, computational biology and nanotechnology. I know that if I start thinking about all of these things that I will start learning how they see the world and what kind of models they use.

I have been interested over the years in distributed systems. If you want to make a complicated system—instead of making this one bulky centralized huge complex system you can often achieve the same functionality or build the system with many little components that are simple and interact with one another. That's an idea that I have been interested in for a long time. The longer that I am involved in research the more I see the commonalities and the more I come back to the same powerful ideas. The same ideas come up in totally unrelated areas such as economics, biology, robotics and electronic commerce.

My approach is about understanding the laws of nature in such a way that we can create artificial complex systems. By studying the complex systems that occur in nature we can understand how to design complex systems that are robust, and adaptive.

Intellectual Renewal   Ben Shneiderman is a prolific researcher and inventor at the University of Maryland[8]:

I'm 54 next month, so I'm aware about the question of whether the greatest innovations are done by people in their twenties. Certainly there are lots of innovations done by people in their twenties but there are lots done by people in their fifties, sixties, and seventies. Part of my goal has been to try to understand how to keep myself in my twenties intellectually.

This is part of the longer theme of creativity, the temporal aspects of it. I am very aware of constantly making myself an amateur or a dilettante and putting myself into new situations. I think that a real danger for aging scientists is when they keep doing the thing that they were successful at.

I deliberately shed certain problems and move onto new ones where I'm not an expert and where I'm not known, where I'm challenged, and where I have a good driving problem.

A model for doing this really came to me explicitly from Gerald Weinberg, the author of *The Psychology of Computer Programming*. He's a very independent-thinking character who I have always respected. He has been influential for me. I taught some courses with him and also went to his farm house in Nebraska for three days just on my own with him. We sat and talked and I remember him making this point about being a dilettante. Every year he enrolls in a seminar for several days on something that he doesn't know about so that he is in touch with what it feels like to be a novice.

This idea has been explicit in my mind—the importance of being a novice. For example, I am now pushing myself into the biomedical field. It seems like a wise thing to do professionally, but it's also the biggest and most difficult thing that I've ever undertaken. One of my pleasures in my work has been to get into different fields and to work with different people in medicine and libraries and legal studies and so on.

The Aha! Moment

Like jazz musicians, inventors gather several distinct kinds of ideas. All the preparation and work of gathering ideas pays off in an exciting way when there is an epiphany or an "Aha!" moment.

What does an "Aha!" feel like? What is the experience of the moment of invention like for inventors who have prepared their minds? Many inventors have told anecdotal stories about this event. We can explore this using a thought experiment.

There is a fable about the invention of chess. Versions of this fable take place variously in India, in China, or in an unspecified exotic kingdom. In one version of the fable, a clever traveler presents the king with the game of chess. The king was so pleased that he asked the traveler to suggest a possible reward for presenting the game. According to the fable, the traveler-inventor said he would like a grain of rice on the first square of the board, double that number of grains on the second square, and so on. The traveler arranged to come back once a day to pick up the next installment of rice.

The king thought that this was a very modest request and asked his servants to supply the rice. One grain of rice was the award on the first day. Two on the second. Four on the third. Eight on the fourth. The amount of rice to be picked up in a week (2^7 or 128 grains) would fit in a teaspoon. In two weeks, however, the daily reward was about a half kilogram. This is about the time that many students get a mini-Aha! When something doubles repeatedly, it grows to enormous size very quickly. By the end of the month, the daily installment would be 35 tons. Although the stories vary on the fate of the king and the traveler, the real lesson of the fable is about the power of doublings. The amount of rice to fulfill the request—$2^{64} - 1$ or about 1.8×10^{19} grains of rice— would by some calculations cover the land mass of the Earth.

This chess fable is well known to students of mathematics and computer science. The story is taught to develop student intuitions about exponential functions and large numbers. Something like this fable crossed the mind of Kary Mullis[9] in 1985 when he invented the polymerase chain reaction. Mullis had a PhD in biochemistry and was working in molecular biology for Cetus, a company that makes synthetic strands of DNA for use by molecular biologists. From his work at Cetus, Mullis was aware of the need for good ways to "amplify" or make multiple copies of DNA. Like geneticists everywhere, he knew about the polymerase enzyme that living cells use to copy DNA. Cycles of heating and cooling are used to trigger each doubling cycle of the polymerase.

According to his account, Mullis was driving through the mountains north of San Francisco with his girlfriend asleep in the car when it hit him.[10] He recognized that he could apply the polymerase reaction to copy a sample of DNA. Then he realized that he could repeat the reaction over and over again, doubling the DNA each time. Then he realized that ten cycles would give him 1,024 copies, twenty cycles would give him a million copies, and so on. Mullis later received the Nobel Prize for his invention of the polymerase chain reaction, now used widely in molecular genetics.

In his description of the event, Mullis writes about how one insight followed the next in a series of exciting revelations and the potent Aha! experience. This rapid sequence of understandings is the same pattern that many other inventors have described.

Marinating in a Problem   Ron Kaplan has a passion for simplicity.[11] Simple theories are better than complicated ones. This principle is called Occam's Razor.[12] Given two competing theories that explain things equally well, the principle says, one should pick the simplest theory. Kaplan:

> I have invented several different things, but the one that gets the most press is finite-state morphology. That has an interesting story, but most of the things that I have done actually have the same character of noticing when things seem more complicated than they ought to be. Why is that? It's offensive when things are too complicated!

Kaplan's invention of finite-state morphology began with a mystery. He was interested in the problem of parsing sentences, that is, finding a way for a computer to determine the syntactic structures of sentences. In his career as a computational linguist, he was familiar with a myriad of possibilities if computers could begin to understand language. He was also deeply aware of many of the simpler problems that lay on the path to achieving so grand a goal.

A first step in parsing is determining which words are nouns, which are verbs, which are prepositions, and so forth. Part of the job is breaking words into their "stems." Stemming enables computers to recognize, for example, that the words 'playing,' 'plays', and 'played' are all forms of 'play'. As Kaplan put it, everybody knew more or less how to do stemming for English:

For English, you strip off the suffixes, take off the "-ed", maybe you add an "-e", take off an "-ing", and then you go and look things up in the dictionary. Everybody knew how to do stemming using suffix stripping algorithms for English and we wanted to go beyond that. We wanted to go and deal not just with languages like English but also Turkish and Finnish and others where you don't just have simple suffixes. In these languages, the vowels change, and from a perspective of the English language, the rules are complex and goofy.

Linguists had studied the different languages of the world, not just English. For about 2,000 years there have been rules that came out of the Sanskrit grammarians. The rules were of a particular kind, what we would call today "context-sensitive rewrite rules."[13] For many years the claim in linguistics has been that these rules were powerful enough to characterize what was going on with umlauting in German, sandhi in Sanskrit, and suffixes in English and so forth. We wanted to figure out how to implement these rules.

Kaplan's search for generality and simplicity led him down a trail first blazed by the Sanskrit grammarians 2,000 years earlier. Kaplan and his group of computational linguists put the linguistic rules into a computer and ran them. Once the linguistic rules were written as computer expressions, they could be tested and measured. When Kaplan and his group looked at the traditional rules for word morphology, they found that they took a long time to run.[14]

This is where Kaplan's desire for simplicity was offended and the need for an invention became apparent. Word morphology—finding the structure of words—was *supposed* to be the *easy* problem. Sentence analysis—finding the structure of sentences—was *supposed* to be the *hard* problem. Surprisingly, the traditional context-sensitive rules for words from the linguists took longer to run on the computer than the rules for sentences:

We worked on and off about 5 years, trying to implement these rules. We could get the context-sensitive rules to work, but the recognition algorithms were more complicated—seriously more complicated, more difficult to understand, and more resource intensive—than the algorithms that we had for doing sentence analysis."

That didn't seem right. You know there's got to be a pony in there somewhere.[15] That was the feeling I had. So *if* it's not really a hard problem, there must be a trick that we're missing.

For any particular context-sensitive rule, Martin Kay and I could get up on the blackboard and work for an hour and a half and figure out a finite-state machine[16] that looked to us like it was doing the right thing. The machine we'd sketch on the blackboard had transitions that would check the context; do the rewriting, and so on. However, there was a problem. The problem was that we couldn't prove that these machines that we were concocting by hand were correct.

The mystery was that for any particular example Kaplan and Kay were able to express the word morphology rules in the simple finite-state grammar. The finite-state machines covered the whiteboard. They knew that in principle finite-state machines were not as powerful as the traditional context-sensitive ones, and they could not prove that their rules were right. This perplexing state of affairs worried Kaplan for about 5 years, on and off. Then as is often the case, the breakthrough came one day when Kaplan was relaxing:

I remember it very clearly. It was a Saturday morning and I was lying in bed and I had the kind of—I don't know whether to call it an "Aha!" experience or an "Oh shit!" experience. [laughs] You know, "Oh shit, this is really trivial!" I realized that you could characterize the particular property under which you implement the context-sensitive rule correctly as a finite-state machine. The finite-state machines were basically enumerating strings. A finite-state machine says "There are some strings that satisfy these properties." In other words, "This exists, and this exists, and so forth." But a context-sensitive rule is a universal quantification, right? It says "For all instances of this situation, do this." My discovery was simply noticing that universal quantification is logically equivalent to not-there exists-not.[17] This meant that the context-sensitive rules were not used at their full power. We could get all the power we actually needed from the finite-state rules.

Several things happened from this experience lying in bed. First I understood that we could get all the power we needed from the finite-state rules. Then I understood why the finite-state machines on the blackboard had been so hard to prove correct—because they tended to be long and complicated. Then I realized that the basic operations we needed to perform to create the finite-state machines were basically set operations like union and intersection. Then I realized that we could make the construction of finite-state machines essentially automatic and provably correct. Then I realized that this made possible a wholesale enterprise of doing word morphology using finite-state machines. Then I realized that we could do word morphology for any language in finite-state machines if we just developed the right linguistic engineering practice.

This was the key to a whole campaign of linguistic engineering. When you combine two finite-state expressions, you get another finite-state expression for recognizing a finite-state language. This is what mathematicians call a closure property. Starting with the initial insight into the logical equivalence, the whole approach for building these systems unfolded before Kaplan's eyes:

The key invention was that you could do algebraic logic over regular mappings. And once you get that idea, then you say "Well, what's the next one and how do you do it?" I had a whole series of insights, one after another. Following the consequences of this insight generated 20 years of people fooling with these things. It enabled us to build these machines. It pointed the way toward the importance of good implementation. So you could build one of these machines that had double exponentiation in it, and it would have, gosh, a hundred states or something in it. No wonder we couldn't do that on the blackboard. You lose track of expressions that are that complicated. So I implemented a number of the basic operations and then arranged to put them together.

I said it's an "Oh shit!" reaction, because once you have the thought, it's so trivial and you think "Why didn't I think of it before?" And then you ask "Why didn't anyone else think of it . . . ?"

At the time of Kaplan and Kay's invention, many companies were looking at the idea of putting word lists into typewriters, text editing systems, or hand-held devices. The main application was spell checking. If a word was in the list, it was assumed to be spelled correctly. In the computing systems of the time, it was not economic to fit large enough lists on a chip. Kaplan and Kay's approach was a bold move. Their key idea was to have a theory of English words rather than a list of English words. Instead of storing an *extensional* representation or explicit list of all of the words, they wanted to have an *intensional* representation or description from which the words could be generated. An intensional description could take advantage of the commonalities of words in a family, so the words 'make', 'makes', and 'making' have commonalities with 'take', 'takes', and 'taking'. An intensional description can be more compact than an extensional one, but building robust intensional descriptions required an engineering theory of finite-state grammars. Kaplan's story is about the journey to the breakthrough that made it possible.

The Use of Challenge Problems   Mark Stefik[18] characterizes himself as a serial inventor, working in one area for 5 or 6 years and then switching to a new interest. The patents and technology from his work on digital property rights were the basis for Xerox and Microsoft launching ContentGuard as a joint venture in 2000.

In the early 1990s, the research-oriented ARPANET was undergoing its transition to the public Internet. Attitudes toward appropriate use of the network were shifting radically. ARPANET was created for government-sponsored research. Any advertising or mention of business was both against the rules and in bad taste. In the 1990s there was a movement to develop electronic commerce on the Internet. "Digital property rights" is a technology designed to enable distributing and selling digital goods such as digitally encoded movies, music, books, and computer games over a network. It is intended to make downloading and purchasing convenient for consumers and secure for authors and publishers.[19] Stefik:

Initially, I didn't have digital property rights in mind as a goal. I had gone to a MacWorld conference with Dan Russell. While we were there we saw PC[20] cards which had just been introduced as a way of extending the functionality of computers, especially laptops. I was fascinated by the idea that these cards could so easily extend a computer's computational capabilities and could contain any kind of processor.

I was looking for something useful to do with PC cards and initially had some pretty bad ideas. Xerox was toying with the idea of putting floppy drives on copiers so that people could copy documents onto them, or print documents from floppies. I wanted to create a visual hand-held device that would show a miniature image of the documents inside. I wanted this device to plug into copiers and also into laptops, and to be easy to use for transporting documents. My idea was seen as impractical because it was so expensive.

What happened at this stage was that Xerox product engineers listened politely and then told me about practical economics.

Electronic commerce had not really started on the Internet. The dotcoms were yet to rise and fall, and the browser wars between Netscape Navigator and Microsoft Explorer were still years in the future. Al Gore, vice-president of the United States, had a reputation for being more technologically savvy than most U.S. politicians. He was interested in helping the Internet to become an important economic force, and he had gathered around him various experts to understand what needed to be done.

In one way, the Internet seemed like a perfect vehicle for electronic commerce. In principle, one could "wire money" on the network and send digital products. However, as the movie studios and music companies were quick to point out, once something was in digital form it was easy and inexpensive to copy the bits. Much of the technical and philosophical discussion about the commerce on the Internet assumed that this problem was fundamentally unsolvable. This issue about the protection of digital intellectual property was the backdrop for Gore's including it on his list of top ten challenges to the research community—challenges that needed to be addressed before the promise of the Internet could be realized. Stefik:

During a brainstorming session we were trying to salvage something to do with PC cards and were also considering what to do with less expensive displays being developed at the research center. We had been playing with concepts for using gyricon or other flat displays on the backs of airline seats so that travelers could choose their own movies. Xerox wasn't interested in that business.

The question arose whether you could store a movie on a PC card. I quickly realized that there wasn't enough storage to do that at the time. Of course, storage density is always increasing. I thought about a second reason not to put movies on PC cards, which was that there was this big issue about making unauthorized copies of movies. I recalled a challenge problem that had been cited by Al Gore, who was talking about ten technical challenges that had to be solved before the Internet could really take off. One of the problems was that there had to be a way of protecting intellectual property in digital form. Copying movies would also be a problem on PC cards. Or was it necessarily so?

And then it hit me. I realized that storage on a PC card could potentially have a very different relationship to the computer than the computer's own disk. On a computer's own disk, *any* program can just read or write *any* data on it. There is nothing to stop a program from copying or modifying any file, and that was by design.

That architectural decision goes back to the beginning of digital computers and their operating systems. It enables computers to be general purpose information processing devices. And it's also why that architecture is so wrong for protecting intellectual property. Arbitrary access and protected access are fundamentally at odds. You can't hold a computer maker liable when some program copies some data because the computer system is designed to be flexible, not protective.

I realized that PC cards could insert a barrier of authentication and control between the processor and particular data. You could have a language describing the transaction rules for using a digital work—such as saying whether it could be copied or not. A program should not just command the disk to deliver up data. It should be trustworthy and follow the rules from the author or publisher associated with the digital work. In principle, it could have a processor in it that checks that certain conditions be met before access is granted. This approach would force computers to speak a protocol and prove that they were trustworthy.

The "aha" experience came as a rapid sequence of insights. First I realized why regular computer architectures and their operating systems were fundamentally flawed for protecting digital works. Then I realized that PC cards could be created to introduce a level of control. Then I realized that the fundamental idea was bigger than PC cards. It was really about a network or community of trusted systems that spoke a secure protocol and enforced rules governing access and use. Digital works could have rules associated with them that said what uses were allowed and computers could be expected to follow the rules. Then I realized that I had found the solution to the challenge problem on Al Gore's list. Then I realized that this was a big deal, because it solved a problem that was generally considered unsolvable. Then I realized that the solution could eventually enable billions of dollars in electronic commerce. It was—whoosh, whoosh, whoosh, whoosh. I didn't come down from that inventive high for weeks.

Working with Ralph Merkle and others over the next few months, additional parts of the invention came into focus—how to handle banking, details of the protocol, how to protect against determined attacks, and so on.

Mark Stefik's preparation for the invention included 30 years of research in computer science. It included his engagement over several months with the properties of PC cards, and his awareness of the list of challenge problems for the Internet. He was searching for the "next big thing" to do. His previous research and his study of computers and their architecture helped him to understand what was possible. His engagements with applications of PC cards and with Gore's challenges helped him to understand what was needed. The alternation between the two questions prepared his mind to recognize the means and ends for digital property rights.

Reflections

Inventors tend to remember their "Aha!" experiences. Every invention involves some small breakthroughs as insights arise to meet the challenges of the problem. Big insights—the Aha! moments—happen much less frequently. Researchers count themselves as lucky if these happen a few times in a lifetime. Ahas are the juice of invention. The excitement of major insights motivates inventors to keep on trying. Even being on the chase of an invention, or what Mihaly Csikszentmihalyi[21] calls "being in the flow," is a strong motivator for focused concentration and invention.

The archetypal journey of the inventor—the movie version—centers on a brilliant individual who has an insightful dream and struggles against great odds before realizing the dream. This myth reflects some truths. Many inventors have moments of keen insight. These moments inspire them.

What is forgotten in the simple versions of the invention myth is that preparation precedes insight. Preparations bring ideas to mind, so they then can be combined rapidly. It is almost like the polymerase chain reaction of the mind. Preparation saturates the mind with ideas. Then when the right problem or the missing link is provided by a situation, the Aha! chain reaction can occur and we have the birth of an invention.

7

The Beginner's Mind and Play

In the beginner's mind there are many possibilities, but in the expert's there are few.
—Shunryu Suzuki

Breakthrough inventions are exactly the ones that *do not* arise incrementally from what has been tried before. They *break* from past experience. In breakthrough inventions, the hard part is finding the insight. Breakthrough inventions require "thinking outside the box." Excitement and buzz permeate a laboratory when there has been a breakthrough. The news travels quickly. ("Did you hear about David Fork's 'claw'?")

Breakthrough or discontinuous innovations are at one end of a spectrum ranging from imitation of proven practice, to incremental innovation, to the risky introduction of novel technology in new applications. Richard Rosenbloom uses this spectrum to analyze the course of innovation in American industry.[1] In Rosenbloom's analysis, a comparative strength of the American system of innovation has been its ability to initiate and rapidly exploit innovative discontinuities that stimulate economic growth by transforming tired industries or creating new ones. Breakthrough inventions create new knowledge and new possibilities. The telephone, the light bulb, radar, and the personal computer are examples of breakthrough inventions.

Stories

Shifting Mindset
A mindset is a set of assumptions that guides our thinking. When we work in an area, we gain experience and acquire particular patterns of

thinking. Over time, these patterns of thinking become deeply ingrained. Without noticing it, we become very efficient at thinking "inside the box." When faced with a novel situation, the built-in assumptions can cause us to overlook inventive possibilities and potential breakthroughs. Shifting mindset helps us to loosen the tenacious hold of these patterns on our thinking.

Shifting mindset is like the beginner's mind. It is about quieting parts of the mind that get in the way of creativity. The beginner's mind is open to fresh perspectives and to unconventional ideas that experts call silly or nonsensical. The beginner's mind is the opposite of trying too hard and getting agitated. The beginner's mind is lighter and more playful.

A saying heard around research laboratories is "I love to give hard problems to graduate students. They solve them because they don't know that the problems are impossible." This parallels a quotation from the San Francisco Zen master Shunryu Suzuki: "In the beginner's mind there are many possibilities, but in the expert's there are few."

Trying the Opposite   When inventors search for solutions, they often start by following their first ideas. Sometimes what they try first turns out to be the opposite of what is needed. Ted Selker recalled an example of trying the opposite when he was working on the TrackPoint invention[2]:

One of the ideas we had along the way for speeding up user interaction was to build in momentum. In the physical world, if you push or throw something in a particular direction, it tends to keep going that way. If a pointing device is going in a direction, we thought it would be likely to continue going in that same direction. We built momentum into the cursor movement routines. In fact somebody had a patent on that.

It turns out that adding momentum to the logic makes a pointing device *harder* to control. In fact, the opposite idea works better. It works better to amplify how movements are *changing* rather than amplifying the current movement. In other words, you take the derivative of motion and amplify the change. When you start moving the pointer it starts faster; and when you slow down, it stops faster. This is the opposite of momentum. The idea is basically ABS brakes[3] on a pointing device. Amplifying the derivative is the invention. The good news is that it makes a fifteen percent improvement in control.

Another example of "trying the opposite" comes from stressy metals in material science. In making of computer chips, one of the steps is the deposition of metal to ultimately create "wires" for circuits in the chips.

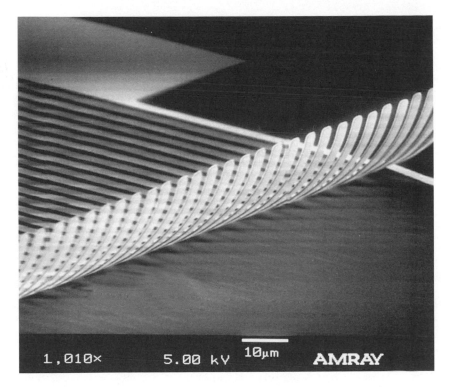

Figure 7.1
A photomicrograph of a "claw." Courtesy of Palo Alto Research Center.

The metal is initially sputtered on to the silicon surface during fabrication to form a metallic film. That step is followed by other processing steps. There is always some stress in the metal, potentially causing it to lift up from the surface. Because such lifting is undesirable for chip operation, the fabrication process is controlled to minimize stress in the film.

David Fork is a material scientist. We interviewed him about two inventions involving stressy metals, one known as "the claw" and one as "kissing springs." In these cases, the inventors saw the potential of going in the opposite direction of conventional wisdom, deliberately using stress to create three-dimensional structures with the metal.[4] Fork:

Don Smith got the idea for the claw when he was writing a chapter in his book about thin film stress and sputtered material. Thin film stress has been the bad guy for many years, and in the minds of many process engineers it still is the bad guy. Conventional wisdom says that if you can't make the stress be zero in your process equipment, then something is wrong. In contrast, the claw invention

Figure 7.2
"Kissing springs." Courtesy of Palo Alto Research Center.

actually relies on there being very large mechanical *stresses engineered* into the material. So it was turning lemons into lemonade in a way to get some three-dimensional self-assembly to occur as a result of there being intentional stresses built into the material ahead of time.

Normally, a lot of bad things can happen when the stress in a metallic film isn't zero. It can crack or delaminate and do other bad things. But it can also result in interesting structures such as microscopic springs or coils. The springs form a bunch of cantilevers that pop up off the surface of a wafer. We call this the "claw."

Here is another example of a structure—a micro inductor that we created by self-assembly. It was processed onto the surface of a CMOS[5] wafer. We constructed it by making a pair of springs that pop up and touch in the middle. We call it the "kissing spring" design. And here is another example of a structure, the brainchild of Chris Chua—a micro inductor that we created by self-assembly.

The "kissing springs" invention can be used to make "sockets" for plugging devices on to a circuit board without conventional ceramic jacks. The microscopic coils can be used as inductors for building high-frequency oscillators on chips. This is useful for miniaturizing wireless devices. Selker's and Fork's examples show how conventional experience can sometimes be turned around. "Trying the opposite" can enable the mind to go beyond the conventional.

The Importance of Play   David Goldberg has strong views about what is hard work and what should count as "wasting time":

It is difficult to come up with good ideas and good projects to work on. That's a problem for managers because they want people to be *doing things*. They need to go to *their* management and say what people are doing. That's bad, because when there's pressure to do something, you just rush in and start doing the first thing you think of. It takes *time* to think of a good project. You need to be relaxed. The time to come up with good ideas is in a different direction than what we normally think of as work.

Once you have a really good idea, of course, then you have to work on it to prove it. But I think people should be spending fifty percent of their time over a period of years not building anything, but figuring out what to build.

Researchers are highly motivated and feel bad if they're not "doing anything." When you're just thinking about what to do, you feel like you're not doing anything. But in fact, you're doing the hardest and most important part, which is figuring out what to do. Later, when you spend a year programming or building your invention, you're not wasting your time on a mundane artifact that has no chance to have an impact. I think that it's much harder to think of a good project than actually to do it.

If I feel that my top priority is to get something done, that's very bad because then I come in every day and try to push all the distractions away. I don't want to go to meetings and talk to people. I just want to get this one thing done. That's really unfortunate because you get the best ideas when you are talking to people and thinking about things. The trick is to have an unpressured atmosphere where people still get things done.

At this point, we asked Goldberg how play fits into his creative life and why there weren't any toys around the office. Goldberg said "Oh, they are over here!" and opened a drawer full of puzzles and toys. Goldberg is known to be an intense worker. In research centers it is not unusual to see creative people juggling or working with puzzles in order to relax.

The Media Lab at MIT often conveys to visitors a certain atmosphere of playful experimentation. For example, there is a two-story space in the Media Lab with offices around the edges on both floors and a large atrium space in the middle. Round tables with computers fill the middle of the first floor in a coffee shop atmosphere. Projectors are available to show things on the walls. Bins of Lego pieces and other toys are on the edges, and yet during certain parts of the day there is a reasonable sense of quiet like a library. Pattie Maes characterizes playfulness as central to the Media Lab culture[6]:

Sometimes I talk about the Media Lab as being like a big sand box for grown-ups. You have all of the tools you could possibly need. You have shovels and rakes and buckets, all of the latest ones, and you know how to use them very well. You just start building things like when kids play in the sand. You start one way without knowing what the end result will be. There is a continuous openness and almost a childlike quality.

Many people have the idea that science is boring. That makes them disinterested in becoming scientists. It is too serious.

It is important that people keep their creativity and passion for playing. I see it in my own children. They are still young. Kids have this passion naturally. They love playing and building things, but somewhere along the way they often become too serious and not as creative and silly. You need to be silly and allow yourself to be naive about things. The student who knows too much and takes everything too seriously cannot be as creative. They should keep a playful attitude.

It is part of the culture in many research groups to encourage an appropriate atmosphere of play. David Fork shared an apt slogan from the Electronic Materials Lab at PARC: "If you aren't having fun, you aren't making progress."

The physicist Richard Feynman was famous for his playfulness. His book *Surely You're Joking, Mr. Feynman* documents some of his exploits. For Feynman, play became essential for cultivating thought. He credits his passion for play as leading to his insights in quantum electrodynamics about how electrons move: "I was in the cafeteria and some guy, fooling around, throws a plate in the air. As the plate went up in the air I saw it wobble, and noticed the red medallion of Cornell on the plate going around. . . . I had nothing to do, so I started to figure out the motion of the rotating plate. . . . The diagrams and the whole business that I got the Nobel Prize for came from that piddling around with the wobbling plate."[7]

It is often observed that the personalities of inventors and other creative people have childlike characteristics, such as a sense of wonder. In breakthroughs, inventors play with ideas and have an ability to put aside "sensible" assumptions in pursuit of creative solutions. This ability may also explain why some inventors are unstoppable punsters. They play with words the same way that they play with ideas. They go over the edge from convention to nonsense. They love to twist ideas into weird perspectives.

Imagination and Creativity   Jacques Hadamard's book *The Psychology of Invention in the Mathematical Field* includes a letter in which Albert Einstein recounts his own creative thinking. Einstein's story is a far cry from "logic." He describes his experience as being mostly visual, and sometimes muscular. For Einstein, the initial formation of ideas is not even closely connected with words, let alone logic: "The words or the language, as they are written or spoken, do not seem to play any role in my mechanism of thought." For Hadamard and others, it is no surprise that words fail to describe this part of the process because it is largely unconscious.

Nick Sheridon described his experience of tapping into his intuition and imagination this way:

Invention is a solitary activity.[8] It's also highly intuitive. Intuitive may not be the right word for it, but it's a highly subjective activity. It's not an analytical process. That side of the brain is not used very much, except to accept or reject ideas produced by the other side of the brain, so to speak. Sometimes you will know that you have the solution but it will take a few weeks to get it to a form where it can be analytically described. Knowing when you have a solution is a gut feeling. It can be frustrating because you can't analyze it yet.

My mother was an artist and painter. When she painted something she had an idea that she wanted to express. She couldn't express it verbally but she could express it in paint. It is very much this way in invention. It's like an artistic process that you can't express in words. Initially, you have to work the words around it. You have to create the words to fit it, and sometimes that process takes a while.

You go through a period of real frustration. The idea is there but you can't pull it out at first. You know the solution is there and that you've solved it. You have the basic concept and now the question is trying to extract that and verbalize it so that those portions of your brain that require verbalization can handle it.

In his work, Sheridon tried to imagine how ink and paper create an image in different lighting conditions. He wanted to create a display that worked like paper in reflecting ambient light. Ordinary paper has no "moving parts," but making a paper-like display requires giving the material an ability to make tiny changes spot by spot. Fundamentally, that meant that the electric paper itself would need to change colors—spot by spot—in order to present different information on a page. Sheridon imagined a set of balls at each pixel that could rotate either to display a black side or a white side.

Putting oneself into such a "pixel-eye view" triggers the imagination. Some inventors have imagined "ink flowing up and down in a tube at each spot." When ink flows up, the spot becomes dark. That line of thinking lead to the invention of electrophoretic displays, in which a material flows up or down in a microscopic tube at each pixel.

This idea of deeply imagining oneself as part of the experimental situation is a recurring theme in accounts of creative invention and discovery. In their book *Sparks of Genius*, Robert and Michèle Root-Bernstein study the accounts of many creative people exercising their imaginations. Einstein, when he was in the early stages of creating his famous relativity theories, imagined himself as a photon racing in the opposite direction of another photon. Barbara McClintock, who won a Nobel Prize later in life for her work on corn genetics, said "I found that the more I worked with them the bigger and bigger [they] got, and when I was really working with them I wasn't outside, I was down there. I was part of the system. I even was able to see the internal parts of the chromosomes—actually everything was there. It surprised me because I actually felt as if I were right down there and these were my friends. . . . You forget yourself."[9]

Joshua Lederberg, who won a Nobel Prize for his work on microbial genetics, said "When I think of a DNA molecule, I have a model of it looking like a rope. And I know, for example, if I pull at the two ends of a DNA molecule, it will break somewhere. Then I have to jump to a different level and say 'Well, I know the structure isn't quite like that.' . . . You have to be able to fantasize in rather crude ways—but then be able to shift from one frame of reference to another."[10]

The creative part of thinking may involve visual or muscular feelings rather than logical or mathematical processes. However, the need to express the insights into a cumulative library of science leads scientists and inventors to report them in a logical fashion *after* they have first developed a sensory or intuitive understanding.

Talking It Out   One of the great joys of working at a research center is having colleagues to talk with. In a great research center, one can often go down the hall and find a world-class expert. Mark Yim describes how this practice works for him:

When I am stuck on something I often take a walk down the hall. Usually I will run into someone and start talking to them. I start to explain the problem. In the process of explaining it, an insight or a new approach may come up. It is rare that a problem has no solution and remains frustrating.

Some people at PARC are especially good sounding boards. Often I would seek out John Lamping. But it doesn't have to be an expert or even someone on the project team. Talking it out works with almost anyone in the lab.

Progress comes episodically. Sometimes we work on something for months without making progress, and then an idea comes up in conversation. We say "Oh, here's an idea. Why don't we try that?" We try it and in three days we have it working fantastically. It may not be along the lines we were thinking before. It may be a completely different approach and one that really works.

The process of explaining a problem or an idea to someone can fundamentally aid in thinking outside the box. Interestingly, this "talking it out" process can work even when the second person is *not an expert* on the problem. The reason is that unconscious assumptions that are hidden to ourselves can surface when one needs to explain a problem to a scientifically minded colleague. In the practice of working alone, unconscious assumptions can remain invisible for long periods of time. When our assumptions are articulated in conversation we sometimes *notice* them for the first time, and that makes it possible to discard or soften them. In general, the less familiar a second person is with our original problem, the more explanations we need to give. In this way, explaining a problem to someone else can help us to cultivate the beginner's mind and enable us to see the problem differently. This phenomenon is probably the basis for the common wisdom in academic circles that a good way to learn about something is to teach a course in it.

Another phenomenon that causes interesting ideas to arise is *creative misunderstanding*. Creative misunderstanding arises when one person explains something to a second person and the second person plays it back wrong. He thinks he has understood, but in fact he has deeply misinterpreted some aspect of the problem. The magic happens when the misunderstanding reveals a solution or reframes the problem.

Getting Away from the Problem   The PARC researcher Diana Smetters has an approach for dealing with situations where the old knowledge isn't working[11]:

Basically I read a lot, especially when I don't know what I ought to do. If I don't think of something right away, I go and read to have more raw materials for new ideas.

I know I need to go away when I find myself in positions where I feel desperate and I am trying too hard. If you can go away from things for a little while you stop trying so hard. Then either your first idea becomes more obvious and rational, or you realize that all the little niggling details that were bothering you aren't actually a problem. Or you may come up with some idea—usually out of what you've been thinking about already. You just have to stop stressing about it enough to actually let yourself think of it.

The language of "letting go" is much like what Shunryu Suzuki called "the beginner's mind." Cultivating the beginner's mind is about letting go of grasping and cultivating spaciousness. Too much alertness—trying too hard—leads to agitation. The beginner's mind has a sense of playfulness, lightness, and receptivity.

But how do you relax when you are trying too hard? Ben Bederson is a prolific inventor in the field of human-computer interactions, especially in zooming user interfaces. A professor at the University of Maryland, Bederson is the second director of the Human-Computer Interaction Laboratory founded by Ben Shneiderman. Reflecting on his experience, Bederson recalled something he noticed early in his research career[12]:

I noticed that 90 percent of my best ideas arose when I was riding my bike over the Brooklyn Bridge. I was in a balanced state of alertness—alert enough to be careful about how I was riding, and not trying to solve my problem of the day. I'd be playing with ideas at the back of my mind. That was when insights tended to arise.

Other inventors take a break or go for a walk to clear their heads. Sometimes inventors, like other creative people, cultivate alternate states of consciousness to see what comes to mind. The PARC research fellow Stu Card has noticed that, as he relaxes just before going to sleep, there is a state of mind in which he can see visual patterns. The patterns arise, fade away, and are then replaced by other patterns in a series of images. This state, known as the *hypnogogic state*, has been correlated with enhanced creativity in various research studies.

Getting Out of Your Own Way   Barbara Stefik has a doctorate in transpersonal psychology. Her clients include many creative people. They may feel stuck in their work, as when a writer is "blocked," or they may sense a need to make changes in their lives.[13] Stefik:

I work with people's dreams with the unconscious aspects of the psyche. There are times when I work with people who are doing creative work but feel blocked, for example, someone writing a book. The ideas are there, but they are unable to access them.

We often go through a process of guided meditation so that the client feels connected to their body. From that place we begin to investigate something that they are interested in exploring. I ask them not to "think" about the answer, but to allow the answer to come spontaneously and to be open and willing to be surprised. Not to have any judgment about what is emerging. Just notice it.

People are often surprised at what comes out, even though it might be something that they feel they have tried to avoid or not reflect on. In the stillness they may come to understand what was puzzling them. With the cultivation of this type of practice, they begin to feel confidence or faith that what they need to understand will be forthcoming. Maybe not in that session, but that it will emerge when it is ready. Then they aren't feeling lost or hopeless, and they aren't completely confused. Sometimes there is an immediate result where people deeply understand something that they haven't understood before.

There are lots of things that can get in our way; things that crop up and stop us from hearing the inner voice and listening deeply. The varieties of fears and blocks that can get in the way are endless. Each person has to investigate and see what it is that seems to get in the way and how to move aside and listen.

Essentially, these are awareness practices. The information *is* there, but if your mind is very busy or more gross it becomes harder to understand what is trying to come through. From a yogic point of view the mind has varying degrees of subtlety. A calm peaceful mind is more subtle. A busy, agitated mind is more gross. If our mind is extremely active, we become disconnected from ourselves. We lose touch with what we are meant to do in life, our calling. When we are more subtle and more contemplative then we can hear more clearly.

Often there is much confusion and polarity—different parts of the self that are at war with one another. That keeps a person feeling confused. It's hard for all of us to hold the paradox of two opposing points of view within ourselves. So we observe that confusion, noticing the different points of view. With the ability to stay open and spacious, and to hold paradox, the new point of view, the unimagined perspective emerges. I think that this different point of view, the one never thought of before, could be a new insight in the case of an inventor.

Guided meditation taps into the truth of the body. The body knows the answers. What needs to be known lies within us. The process of guided meditation is a process of connecting with the body, breathing into the body and really settling into the body. It's amazing how often we are in our heads and disconnected from our bodies, especially our hearts. So the guided meditation is about body awareness and settling into our bodies at a more and more subtle level.

If you are talking about a specific invention or something you need to solve and you don't know the answer, again you can go within yourself and just relax without having any preconceived ideas. You let go of any ideas of how it should look or what it should be. You wait quietly. If nothing is there, then nothing is there. I'll say to a client "OK, so what is 'nothing' like?"

"Oh, it feels a little tight in my chest."

"Can you go to that tightness? What is there?"

It might open up more. There might be some more subtle constrictions that are in the way of what's trying to come forth. So you keep working more and more subtly to see what emerges. I tell a client to tune in and to listen deeply and reflectively. Be willing to be surprised and don't be attached to what's coming forth.

What emerges might not look like what you think it should look like. The deadly piece is that we keep holding on to how we think it should unfold, and what we think it should look like, while we pull on familiar techniques that have helped us in the past. If none of these are working in this case, we're going to feel lost or stuck. The new thing is often very unique and surprising and not at all what we thought.

When people meet up with the insight, they say "Aha. I never thought it would unfold this way. In hindsight, it makes perfect sense. But if you had asked, in a million years I would never have figured it this way."

There's often that really surprising quality to it which is very fun. Some people who are good at this can learn to enjoy that quite a bit. "Wow, I wonder what will unfold this time, because the truth is I don't really know." There's a lot of humility, throwing up your hands and saying "I don't know but I'm willing and open to what may come."

Systematic Search Approaches

Systematic searches are programmatic and generally combinatorial ways to search for solutions. For example, we might draw a table of alternative solutions with rows and columns to characterize and compare them. Systematic search encourages a discipline of writing down the dimensions of different solutions and then generating more alternatives in an exhaustive and combinatorial manner. The "good" ideas that arise in a systematic search are usually few in number and hidden in the midst of many "bad" ideas. Systematic search can help us to find breakthroughs if we are alert enough to notice them.

Failing Quickly and Often   Gary Starkweather, who in the late 1970s invented the laser printer at Xerox PARC, is now a researcher at Microsoft. He characterizes effective research as learning to fail quickly: "The secret of research is learning how to fail fast. That gets you to the right answers quickly. You measure good research by how fast people are making mistakes. It's the one organization in a company whose job is to explore."[14] Starkweather's dictum emphasizes the rate of exploration. Most new ideas are bad. From this searching perspective, a strategy for research is to search alternatives quickly, without fear of failure.

Starkweather's story is similar to an anecdote told about Linus Pauling. Pauling was asked by a student "How does one go about having good ideas?" Pauling replied "You have a lot of ideas and throw away the bad ones."

Learning to fail early and often is a tenet of industrial design and a theme practiced in the Center for Design Research at Stanford University. Larry Leifer[15]:

In design we tend to use the phrase "fail early and often" to describe the method of generating many ideas or concepts early in a product design or development cycle. This helps to make later stages relatively error free. Even in structured models of design through phases, we move on to the next version quickly—again "failing early and often." In this way that attitude dominates our design practice.

In some of our principled research studies of design practice we have found evidence that failing quickly actually matters. These were double-blind studies to judge the quality of a design outcome and correlate that with the design practice. We measured the design quality with an external panel that did not know which team did the designs. We measured design practice using an analysis of a design transcript. We used the rate of adding noun-phrases as a stand-in for introducing new design concepts. In other words, we counted the number of noun phrases that people used as a measure off the number of ideas that they generated. We also looked at the rate of asking certain kinds of questions, which is a stand-in for criticism and the number of ideas that they rejected. Groups that "failed early and often" generated and rejected more ideas than the others. The results of the experiment showed that these metrics were a good predictor of the design quality. The more rapidly a group generates and discards ideas, the better the resulting quality of their design.

The virtue of failing quickly is counterintuitive for people who expect designs to unfold directly from a principled linear process. The practice requires letting go of perfectionism, and developing confidence in a process of trial and error.

Seven Bazillion Subproblems   Diana Smetters has worked in experimental neuroscience, cryptography, and security. Her current research deals with user-friendly security designs for mobile devices. The sign outside her office door says "CAUTION: Dangerous when bored." Smetters:

I started thinking about security for wireless devices. The big problem with wireless technology is that previously you think about the walls of your house as providing you with protection and limiting what people can see and do. For wireless devices the distance that they radiate is not limited by the physical structures that we normally think about providing security.

If you walk on the street or ride on the subway, you may want to have some kind of anonymity and privacy. Even for the devices that stay in your home, how do you put the "walls" back?

Because I work in security, I usually pick things that I think are interesting based on what scares me. I first started thinking about this when I was reading about Bluetooth. Bluetooth is a local, ad hoc wireless technology. It has pretty heinous security like most things created by industry groups.

There is some really bad brokenness about it. Its minimum range is ten meters. I think it was invented by people who live in large ranch houses in the suburbs, where ten meters away means definitely out of range of any bad guys. It probably means inside your house. And when I first read about this I was living in Manhattan. People who were ten meters away from my apartment were on the other side of razor wire. The idea of having something that radiated in a circle that was ten meters wide made me very nervous. Anyone in that circle might be able to see that I'm doing something even if they can't take advantage of my devices. My neighbors downstairs in the next building were the ones on the other side of the razor wire. Within ten meters of me you could cover basically my entire nine-digit ZIP code, which turned out to cover a floor and a half of my apartment building. So generally I started out not being thrilled.

I was at the same time trying to follow up on Tom Berson's idea that if you give people more cryptography, they will come. In my opinion, we had come to the end of our Quicksilver project for cheap, accessible security operations. The problem was that cryptography is not an intrinsic motivator. You give people more cryptography and they won't come. You don't buy a burglar alarm until your neighbor's house is broken into.

I started working on this wireless security project with Dirk Balfanz and Paul Stewart. We assumed a future where you live in a world with lots and lots of wireless devices that are not terribly sophisticated; don't do a lot of functions on their own, and where a person has lots of devices that are never seen by a system administrator. People have to manage all on their own. You can't explain to them how public-key encryption works because most people don't want to know.

The question is: If usability is your overwhelming concern, how much security can you get? If you want people to actually use the stuff and not just turn the security off, because it's complex and bothersome, can you give them enough security to protect them? We assumed that the threats were low level, not from well-funded governments or something like that.[16]

Can you give them a model for how you might securely connect two devices? Or introduce two devices to each other and then have the devices to have enough smarts to manage their ongoing interactions. The idea is to put a little bit of effort into your security up front, by "introducing" new devices to your home system. The home environment is smart enough to smooth over continuing security interactions from that point onwards.

Once you have solved the top-level problem, the challenge is that there were seven bazillion problems below. These subproblems were about how you solve parts of the big problem and how you implement them. Certainly most of what's come from that is a list of problems that we haven't solved yet. And then there's all these other things—"How can we actually implement this?" and "How do you deal with having multiple groups of devices?" and "How do you cross groups of users?" and "How do you use credentials for one group to talk to things in other groups?"

So you've got a running inventory of problems here. You know which subproblems are solved and which ones are open. The solutions are going to come piecemeal and you might not get an "Aha!" for the whole thing because you still have these details that might be bothering you.

You need to sit with a problem and play with it and perhaps build some systems to get a grasp of the whole thing. At this point we are doing a lot of thought experiments and starting to build some prototype systems.

The challenge in this case is about dividing and conquering. When Smetters talks about "seven bazillion problems," she is tackling the big problem of designing usable security for mobile devices. The challenge is to identify and orchestrate the smaller, tractable parts of the problem. The breakthroughs in this project arose when Smetters and her colleagues redefined the problem away from the conventional approaches, which require complex "public-key infrastructures" and are difficult to use.

When Smetters talks about cryptographic research and groups of users, she is implicitly crossing field boundaries on a research project that followed a problem to its root. The original formulation of this research identified a problem of security in systems networked by wireless networks. At first the problem seemed to be the absence of a public-key infrastructure (that is, an infrastructure for registering users and assigning them unique cryptographic keys). Several companies have tackled the problem in that way, and all of the known solutions have been complex in that they required network administrators. However, network administrators represent technically savvy and expensive labor at a level that has been difficult to use in small offices or in homes. This revealed a new obstacle: if security systems are too difficult to use, they will be ignored. The security group invited PARC social scientists to work with them and began identifying a different approach. Following the problem to the root led to new technical approaches that emphasized ease of use for appropriate levels of security.

Generating and Testing The generate-and-test approach to problem solving combines a systematic way of generating candidate solutions with an efficient way of throwing out the bad ones. The assumption behind the method is that it is easy to recognize good solutions. The hard part is to generate candidates.

When Mark Stefik was a graduate student in computer science at Stanford, he worked on a computer program for figuring out the segment structure (piece structure) of DNA molecules. One afternoon he had two encounters with the generate-and-test approach in problem solving that left a lasting impression.[17] Stefik:

At that time I was a second-year graduate student in computer science. Preparing for my thesis, I was spending a lot of time in Joshua Lederberg's genetics laboratory in the medical school. Researchers in Josh's lab were always conducting experiments where they were digesting DNA, that is, cutting DNA into fragments using enzymes. Jerry Feitelson and I came up with the idea of writing a computer program to infer the DNA structure from the data.

Researchers in the lab were doing enzyme digests all of the time. Figuring out the segment structure was reasonably difficult, so people might spend hours doing it for each experiment. We thought that a good first project for me would be to automate the reasoning for these problems.

I started out by interviewing people in the lab about how they solved the problems. There was an underlying set of rules that governed how the DNA molecules were split apart by enzymes. My program was supposed to look at the digestion data and figure out how the segments were arranged in the original molecule. In any given experiment, a molecule might be digested by more than one enzyme. There were also experimental techniques called "partial digests" where the enzymatic processes were interrupted before they ran to completion. The partial digests gave intermediate-sized segments, which were clues about which pieces were next to each other.

At first things seemed to go quite easily. I interviewed people, wrote down their rules of reasoning, and programmed them in the computer. The program would re-assemble representations of the segments into representations of larger molecules. But as I explored more data, something began to bother me about the way the program worked. The more data the program had—the more clues it had about the segment structure—the longer it took. This seemed completely wrong to me. A puzzle should get easier when you have more clues. I set up an appointment to ask Lederberg for help.

I sweated for two weeks. I have tremendous admiration for Josh. In my mind he was high on a pedestal. I did not want to go and ask him a stupid question. Finally, on the afternoon of our meeting an hour before our appointment, I figured it out. I needed to reorganize the program so that it used the enzyme data as constraints or data for pruning rules rather than as clues for assembly. Having Josh tell me this *would* have been embarrassing since he had pioneered the Dendral algorithms that used this approach for organic chemistry problems.

There was actually a saving grace to the interview. After staring at these problems for so long I came upon a new kind of experiment for elucidating DNA structure and was eager to tell Josh about it. I thought that it would make it easier to solve more complex structure problems than the ones being done in his laboratory. My idea was to do a partial digest with one enzyme, and then a complete digest in the other direction with another one. I was pretty proud of my idea.

When I met with Josh and told him my idea, the most amazing thing happened. He scratched his chin for a moment and said "That's interesting." Then, as I remember it, he went to a whiteboard and started listing the variables we could control: one, two, or three enzymes, complete digests versus partial digests, two and conceivably even three dimensional arrangements of the gels, and the order in which the digests were performed. He took my single variation on their experiments and in two or three minutes generated a whole family of additional experimental variations. I was dumfounded.

Josh showed me a *new way to think* about problems. You could call what he did a kind of dimensional analysis. This left a lasting impression on me. Now when I get stuck I look for ways to generate variations. I will always remember that afternoon.

Lederberg's demonstration for coming up with *kinds of experiments* was yet another variation of the generate-and-test method. He used the experimental dimensions as a basis for *generating* new kinds of segmentation experiments. The deep lesson for Stefik was that it is often possible to use a *systematic search* for candidate solutions in many kinds of problem solving.

The power of the generate-and-test method is dramatic when it is possible to create a complete generator—that is, one that will create without fail all possible solutions. This was true for the DNA-segmentation program that Stefik created, and the earlier Dendral[18] systems for organic chemistry created by Lederberg, Djerassi, Feigenbaum, Buchanan, and others at Stanford.

Carl Djerassi, then a professor of chemistry at Stanford, enlisted his graduate students to use the later generations of Dendral to check results published in a chemistry journal for completeness and correctness.[19] He asked each student to pick an article on the structure of some natural product based on a variety of physical methods. He then had the students check the results, entering the published information into the computer system. As Djerassi reported, the results were dramatic: "Without exception, each student discovered that the evidence cited in the published literature was consistent with at least one structural alternative that the authors had not considered." In other words, the chemists routinely made mistakes and missed answers. Apparently far ahead of his time, Djerassi suggested to the editors that they include such computer review of articles in the review process for their publication.

The use of Dendral by Djerassi's graduate students demonstrates that structure-analysis problems in chemistry have become too complex for reliable reporting in the scientific literature. The kinds of errors that

Djerassi's students found routinely escape the attention of the authors and their reviewers. As Lederberg has noted,[20] a significant part of the chemical literature has now become shaky. The underlying difficulty is that the scientists used unjustified assumptions in reporting their results. Avoiding such reasoning errors is quite difficult for people when the problems are very complex.

Reflections

The Prepared Mind versus the Beginner's Mind

From a certain perspective, the "prepared mind" and the "beginner's mind" seem to be opposites. The prepared mind says "Use your previous experience"; the beginner's mind says "Discard your previous experience." One way to reconcile the seemingly contradictory advice is to recognize the difference between routine and novel problems. The advice about having a prepared mind is for routine problems; the advice about cultivating a beginner's mind is for novel problems. For routine problems, our experience serves us well and helps us to work through the problems more effectively. For novel situations, however, parts of our experience can get in the way of having a breakthrough.

For inventors that face a mixture of routine and novel problems, the methods of the prepared mind and the beginner's mind are the yin and yang of creative problem solving. An inventor working on a problem starts out by trying to apply knowledge from past experience. If the old experience doesn't seem to work, the inventor will face an impasse: "I've never seen a problem quite like this one before." "I can see a couple of alternatives, but neither one quite works and I don't know what to do." The impasse is the signal to cultivate the beginner's mind. When old knowledge doesn't work on new problems, cultivating the beginner's mind helps free inventors from preconceived notions. Inventors can notice and put aside "sensible" assumptions to find creative solutions.

From a philosophical perspective, learning how to cultivate the beginner's mind can also be understood as a variation on preparing the mind, rather than as its opposite. When we learn to try the opposite, relax, or carry out a combinatorial search, we are learning something about *how to approach novel problems*. People who study problem solving sometimes characterize the methods of "the beginner's mind" as problem solving at the meta-level. In other words, cultivating the beginner's mind prepares the mind to break out of mindsets.

Balancing Play and Work

Play involves both deliberate search and movement with no apparent goal. Playing with ideas gives inventors safe ways to sneak past their censors, enlisting the wisdom of the unconscious. If you are bored, you are not engaging your full capabilities. In their book *Sparks of Genius*, Robert and Michèle Root-Bernstein collected and analyzed accounts about how people cultivate creativity. In one example after another, they found scientists, artists, and other creative people incorporating play into their work lives. One of their examples was Alexander Fleming, the discoverer of penicillin. Fleming's colleagues knew that he liked to play. He always made time for sports and games. He also enjoyed changing the rules to make games more difficult. For example, he might play a round of golf using just one club. Fleming characterized his professional work as playing with microbes. One of his seemingly useless pastimes with microbes was bacterial painting: he would brush variously pigmented bacteria onto agar plates,[21] and when they bloomed they would producing pleasing pictures.

Fleming apparently kidded his lab partner, V. D. Allison, about excessive tidiness. Allison would put his lab bench in order at the end of every day and discard things for which he had no further use. Fleming , in contrast, would keep cultures for two or three weeks and would look at them carefully before throwing them away. This practice led to the accidental discovery of *Penicillium*.

Fleming had much in common with Max Delbrück, one of the founders of molecular biology. Delbrück had a "principle of limited sloppiness": Be sloppy enough that something unexpected may happen, but not so sloppy that you can't tell what it is.[22]

Sometimes companies face dire situations, such as when revenue is down and survival is at stake. Management can create a sense of urgency when it becomes consumed with "making reports that justify what we're doing." Such urgent activities compete for time and mind share with activities that cultivate creativity: "This is no time to play! Don't you realize what is at stake?" The flip side is that sustained hard work without play and inspiration can be a path to boredom and mediocrity.

Research and invention are not about playing all the time. Invention has a rhythm. Play is most crucial when creativity is most crucial. Without play, the joy drains out of the work. It becomes harder to hear the inner voice and to engage in novel ways of thinking. For many first-rate scientists, play is an essential part of their method. Repeat inventors create a balance, sometimes working very hard and sometimes playing, finding a rhythm that serves both the routine and the creative parts of their work.

8

Walking in the Dark

It was already late
enough, and a wild night,
and the road full of fallen
branches and stones.
But little by little,
As you left their voices behind,
the stars began to burn
through the sheets of clouds
and there was a new voice
which you slowly
recognized as your own,
that kept you company
as you strode deeper and deeper
into the world.
—from "The Journey," by Mary Oliver

The feeling of being lost is known to all repeat inventors. Getting lost is a recurring experience and a normal part of the creative process. When a journey is long enough, getting lost is likely. Feelings of inadequacy, doubt, and confusion can arise when an inventor gets lost. "Perhaps there is no solution," an inventor might worry. "Perhaps I am working on the wrong problem. Maybe I should try something else."

The frustration and confusion of a dry spell can be very disheartening. The good news is that there are ways to rekindle inspiration and get back on track. Repeat inventors learn these ways and aren't as consumed by the fear and confusion of getting lost. They may not know where a solution is, but they have learned where to look. They develop skills for "walking in the dark." As inventors and researchers gain skill and confidence, they often take on harder problems. The stories that follow are about how inventors handle being lost, and how they handle the darkness when ideas seem to dry up.

Stories

First Experiences in the Dark

Henry Baird is well known for his research on statistical image analysis,[1] optical character recognition, and pattern recognition. He has worked on many inventions at Bell Labs, at RCA's David Sarnoff Research Center, and at PARC. He recalled his formative and early struggles when he was a graduate student at Rutgers University[2]:

I remember when I was writing my masters thesis, which was the invention of what is now called the sweep algorithm. This is a fundamental technique in computational geometry and is also widely used in the CAD[3] artwork verification industry. I discovered all of this while I was writing my masters thesis.

I was feeling very insecure, unskilled, and very much alone—sitting at my tiny desk and trying to crack this problem. I was trying to force myself into believing that the problem was important and that I might come up with a useful answer to it. None of it was clear—there was a huge lack of clarity. Am I working on the right problem? Perhaps not. Maybe I should abandon it and go to another problem? I knew that if I went on to another problem that I would be assaulted by the same fear. It really takes a lot of moral strength to continue working on something even through it may take months or years to know if you're working on the right problem. Persistence is very important. The ability to tolerate loneliness is a critical determinant for success.

There's no one you can turn to. You and the problem are in this mortal struggle and you don't know if it's the right problem or you'll get a good answer to it. And you just have to have faith and to keep on pushing.

In pushing on, you have all kinds of feelings. I don't know what it is that allows people to hold their finger in the flame as long as they do. But I wanted to be a researcher. I wanted to be a creative intellect. I wanted that to be my self-image. And I desired it very strongly. I went to school nights and I choose the masters thesis option rather than the essay option, the easier way out. I took the hard way at every step, because it would lead to the best results. I wanted to be very, very good at something. But there was no guarantee that I ever would be. There was a risk. And so here I was betting several years of graduate study and months of work in isolation that I could achieve something.

One of the ironies is that after I finished it, I had no idea how good it was until it was published. When I first gave a talk on my thesis, I was very insecure and felt I was wasting people's time. Then it won the best paper award at a conference. There was this shock—this transforming metamorphosis—that reward has on young people in research. That is why it's very important to recognize the work of young people, to praise them, and to reward them. Otherwise they may never recognize how good they are. Very often a young person is suffering from low self-esteem or is uncertain, and therefore does not recognize how good the work is. That has to be drummed into them later. Publication is very important because it's the only way you can get a response from a large audience.

Freedom is important in the life of an inventor. If you are a good researcher, you're at the cutting edge of your field. No one can tell you what is the right direction to go.

There are very few careers in which a person can be rewarded for being good at something as opposed to being useful to others. I think research is one of those careers, where the utility of research is so far off in the future it may never be revealed in the lifetime of the researcher. So it's an opportunity for a person to achieve something that is good in itself. I think that's a kind of pure joy that mathematicians have readily and steadily in their careers, and scientists and inventors can have, and probably creative artists as well. I think it's the same joy in all of those fields. And I think it's rather hard to find it outside of the arts and research and mathematics and so forth.

I should also say that most of the best researchers work to a large extent in isolation. You are constantly betting your career on your instincts, and that can be lonely.

Although Baird struggled, he was determined to develop himself as a researcher.

Heading for the Data

Stuart Card's expertise is in understanding and inventing effective human-machine interfaces. With the wrong designs, these interfaces can lead to poor and error-prone human performance. Card is often asked to take on difficult challenges and he knows that he will sometimes get lost. This happened with high stakes when he was asked to help understand some problems in the controls of new fighter planes. In this story, Card investigated what was going wrong and what was needed.

The cockpits of aircraft, especially helicopters and fighter planes, are among the most extreme environments for human-machine interactions. On several occasions, Card has been asked to participate in studies of human interfaces in these environments for the Air Force, the National Research Council, and the Army. In each case, the study has been commissioned because a problem has come up in pilot or aircraft performance. Most of the participants in the study are from the military or aircraft industry and work on the specification, design, or manufacturing of planes. The situation may involve high stakes, depending on the findings of the study. Usually, there is a level of confusion (and potentially finger-pointing) as the study commission tries to ascertain what is going wrong.

When Card began his participation, one of the problems he faced was that the planes were new and information about them was highly restricted. He found himself under pressure, trying to figure out the cause of the poor performance[4]:

I was heading the human-factors part of a major Air Force summer study. My counterparts—one was a former chief scientist of the Air Force, the other was a vice president of McDonnell Douglas—and then there was me. I had the behavioral side. All these other guys were coming up with important findings, but our committee wasn't doing so well.

Several times when I've gotten in trouble, I can get out of trouble if I go back to examining the behavior. So I tried to follow that principle. I had them fly me in some test pilots. I got two Viet Nam fighter pilots. Then I got the chief test pilot of the Air Force. They made him fly an F-16 jet all the way from Nellis Air Force Base so that I could go over it.

When I got the pilots, I had them describe to me what they did in combat. They told me stories. I actually had them give me some time constants. I discovered that they operate on a cycle of about five seconds per move.[5] They make their move and they're thinking of the next move before they've made the last one. That's a lot like chess. For a Grand Master chess goes at five seconds a move.[6] Not only that, but if I add together their flying times, I discover that they're up in the 10,000–15,000-hour range. In other words, they have about the same amount of hours of practice as a chess Grand Master. Chess is partly a perceptual-cognitive skill. You can see that, because a Grand Master makes a move in five seconds [clicks finger] and can play twenty games at once, going down, looking at the board, and moving in five seconds and going back. I knew that because the guy that did all the chess studies [Hans Berliner] was my roommate in graduate school. I used to interview him by the hour about this stuff. There is a perceptual basis to this skill. And hence, a lot of things of the chess literature come over to fighter pilot situation. The timing tells you about a lot of things, such as about chunking.[7]

The study of human performance at chess has a long history in cognitive psychology. Card's interest in the "five seconds per move" phenomenon ties deeply into theories about how a chess Grand Master (or an expert in some other task) learns to quickly recognize patterns in situations and then react to them with high skill. Recognizing the five-second loop in the behavior of a pilot in a fighting situation, Card was finding his way into the perceptual and cognitive parameters of the task:

The key to understanding was getting in there and seeing, drowning in data, really sitting down with it. If you just sit down with the data and pore over it in minute detail, an idea will occur to you. That's my basic method of protocol analysis.[8] If you can put in enough hours, an idea will occur to you.

We had to do a brief-out in front of thirty Air Force generals and these guys do not suffer fools gladly. Ultimately our brief-out was the most successful. The key was going down to the detailed behavior, getting enough data. I made them fly in jets so that I could sit in the cockpit and see the displays and get enough contact with the actual experience.

Later, on the same study, they flew in an F-18 when there were only twenty in the country. A hapless marine test pilot who was testing it down in Virginia somewhere had to fly it up for me. We went up to the local National Guard base. When the F-18 landed, it was the first one anybody at the air base had seen. A couple of minutes later a little Cessna landed with a ground crew. It flew in from McDonnell Douglas in order to service the F-18 because these guys wouldn't know how to service it. They parked it at the end of the runway. Then they wheeled up a big generator, fired up the generator, and put an umbilical in it so that I could climb up in the cockpit and see the heads-up displays without actually flying up in the aircraft. That was important, because without seeing the displays I couldn't understand how complicated the joystick was, given everything a pilot had to do. I needed to see what the display brightness was like so that I could interpret the stories. The pilots would tell me that "this and this and this happened" and now I could envision it.

The same thing happened for another study that we did on modeling to build the LHX helicopter. Jerry Elkind and I had a helicopter flown down to NASA. They flew it all around and did all these maneuvers for us. Out of that I got the information that we needed to understand the basic problem.

In these examples, Card walks into the situation cold and needs to figure out what to do. Contributing factors could be anything from pilot training to system design, and the answer is different each time. It could be any of a number of factors, including cognitive overload, bad display design, delays in responsiveness, or even a poorly designed joystick. Working from the initial reports would not give Card a basis for figuring out what was going wrong. He recognizes that his *own* expertise can emerge if he immerses himself in the behavioral situation (the actual cockpit, the pilot's stories, and the plane's controls and instrumentation) and in performance data.

Trying Too Hard

A common tendency when something isn't going easily is to try harder. We all do this. If the wheelbarrow won't budge, push harder. However, in invention trying harder is not always the best way through the darkness.

Nick Sheridon has led and contributed to a large number of inventions including acoustic holography,[9] ruticons (image-recording devices), improvements to diffraction gratings, and early multi-functional (fax/printer/copier) desktop devices. He is often working on several projects at once. Like other repeat inventors, he knows that the initial insight is only a beginning. Unforeseen problems arise along the way for which solutions may not yet be known. In Nick's work on electric paper, no robust theory had yet been developed for many aspects of the invention. Problems in cavity formation (for the balls to spin), adhesion, temperature sensitivity, ball-size control, and sheet curing arose one after another.

When a problem doesn't yield to a quick engineering fix, Sheridon often finds himself stymied, with no specialized book or theory to turn to. Sheridon has become familiar with the feeling of coming to a dead end and is reflective about what to do when this happens[10]:

Sometimes you will work on a project and you will come up to a dead end, at least temporarily, because you have run out of ideas to try. The problem is you are trying too hard. If you're trying too hard, then you loose your openness. You tend to fall into *fixed patterns* of generating ideas—more analytical and less open. By openness I mean really open. You've got to let go of every concept. It's a willingness to accept; a willingness to know that you don't know; a willingness to recognize your complete ignorance. That's all part of it. Humility is a very important part of it. You could get into theology and say it's self-emptying. It is probably related to that in some way.

If you've got all of the information that you need then you are better off staying with logic. You occasionally get into a situation where you really don't have enough information to make a logical choice. You try, but there's not enough information to base anything on logic. Then you do go into intuition, at least to listen to it and see what it has to say. I often make a list of all of the properties that I want some invention to have when it's done.

In making a list you start with the logical process of trying this idea and that idea, and then fitting them against the list. At some point there's a stroke of genius or a flash of inspiration. The patent office loves to talk about "strokes of genius." That's where the term comes from. In a way that can't be explained really, the concept about what you want to do comes up. Often the concept can't be verbalized. It's a feeling. You've got it there and you know the solution is there. Sometimes it will take weeks to get from there and you're not really inventing during that process. You're trying to verbalize the concept.

Sometimes the work takes place on a long drive or in the shower. It is subconscious. That is what is happening. It depends on the problem though. Some problems are easy and you don't go through all of this. The more fundamental—the more difficult problems—that's when it occurs. You're basically opening yourself up to it.

Nick Sheridon has worked on many inventions and has gotten lost many times. For him, getting lost precipitates a familiar response: "Ah yes, here I am, lost again. This is a dead end. I am probably trying too hard. I need to wait for a solution to surface." The solution can't be rushed.

Reflections

Getting Past "Genius"

First encounters with flashes of insight can be heady experiences. An insight builds enthusiasm and confidence. It is wonderful to enjoy and cherish an insight. However, after the flush and the enjoyment, attributing the insight to personal "genius" can be problematic.

Pride and strong identification with an insight set the stage for later hazards. When obstacles appear, strong identification with the insight leads to *personal* doubts. "Was *I* wrong? If *I* am so smart, why am *I* lost?" Everything seems to be at stake. Confidence comes tumbling down.

For repeat inventors and experienced researchers, confusion and fear of getting lost become less personalized. With practice, repeat inventors become less attached to the belief that ideas flow from their own genius. Instead, they become more interested in the *conditions* under which ideas seem to arise.

Creating Conditions Under Which Ideas Arise

Earlier we identified the four main ways to discovery and invention: need-driven, data-driven, method-driven, and theory-driven. These four ways can help the lost inventor, since they each offer an approach for rekindling the spark of invention.

In the story about his participation in studies of aircraft controls, Stu Card reflects on what he does when he gets in trouble. To Card, waiting for ideas is not useful. "Don't just sit there, do something!" But there is more here than simply a bias toward action. Some actions work better than others for bringing about insight. Card's approach is to gather and study data.

Card listens to the pilot's stories, experiments with the cockpit controls, and makes graphs showing how long it takes to carry out various actions. His method is to marinate in the data, knowing that *something* will turn up that will show him what is going wrong. In effect, he knows how to look at the data for patterns that will be clues. Card's approach in this story reflects the data-driven approach.

The need-driven approach starts with a problem and looks for solutions. New inspiration using a need-driven approach may arise by searching the world for elements that satisfy parts of the problems. Another idea related to this approach is to re-examine the problem being addressed. Perhaps the problem has some unnecessary assumptions that could be changed. Changing the assumptions may change the problem just enough to allow a fresh solution.

The method-driven approach uses advantaged instrumentation to gather data. New inspiration may arise by making it possible to see or measure things that are currently invisible by changing the instrumentation. Perhaps a new "telescope" or "microscope" suitable for the problem would help a researcher to understand what is possible.

The theory-driven approach uses theories to give advantaged ways of thinking about a problem. New inspiration can arise if one tries a different kind of theory. For example, a researcher could look for extreme cases in the current theory. Maybe an analogous problem has arisen in other fields. Sometimes an analogy from another field offers a clue to finding an unexpected solution.

All four ways help inventors to cope with being lost. They potentially lead from the darkness and confusion to greater clarity and fresh ideas.

The Right Level of Effort

The strategies may seem contradictory. In Henry Baird's story, as a graduate student, he is concerned about the isolation and doubts that arise and how determination laid the groundwork for his life as a researcher. Baird recognizes that persistence is essential and wonders "What allows people to hold their finger in the flame as long as they do?" In recent years he has become a deliberate networker, connecting with colleagues and collecting their ideas.

Nick Sheridon understands that trying harder doesn't always lead to solutions. The meditation instructor Sylvia Boorstein plays with the familiar admonition by twisting it in the title of her book *Don't Just Do Something, Sit There!* Getting busy without reflection does not yield the necessary capacity for change. Being lost becomes a signal for a shift in strategy rather than the beginning of fear.

Apparently a balance is needed. Repeat inventors learn to "try hard" and also to "not try too hard." This balance involves awareness and is learned experientially. It requires having enough persistence and patience to work through difficulties, and enough awareness to recognize when it is necessary to restore energies and be open to shifting approaches.

III

Fostering Innovation

9

Innovative Research Groups

When we first started our courses on innovation and design, we thought that designers had to be Renaissance men—mastering all of the subjects to do whatever was needed on a project. It became painfully clear that people couldn't do that. Only one student in a hundred could do that. We switched our approach to creating and coaching *design teams*.

—Larry Leifer

Interactions with researchers and others can stimulate the creation of ideas. Researchers and inventors often prefer to work in research groups. A sufficiently large research group provides a *critical mass* of intellectual support. (Too small a group may not provide enough stimulation or enough coverage.) Projects that are too big for individuals to pursue effectively can be tackled by teams, especially in engineering.

Stories

Creating Innovative Groups
All research organizations want to hire and keep the best and the brightest people. Research managers and department heads know that effective hiring is essential in building a world-class organization. However, hiring great people by itself does not make an organization innovative. Without collaboration, there is little leverage. The research group becomes only loosely connected. Without synergy, the overall productivity of a research group is no better than it would be if the researchers worked in separate organizations.

Hiring Practices Bob Taylor is highly esteemed for his success in getting people to work together on the shared dream of using computers to help people communicate. For many years he was a research manager at the Digital Equipment Corporation. Before that, he headed Xerox PARC's Computer Science Laboratory (CSL) when the personal computer was being developed there.[1] Taylor:

I'm not sure whether to call it invention, research, or engineering. Let's just say that the processes include all of those things.

You begin with someone having a dream—something that they want to accomplish. I don't mean in terms of a personal accomplishment, a salary objective, or a positional appointment. No. Rather, they want to effect some change in the world. My guess is that all the best inventors begin with that kind of stimulus, coming from within them.

The second thing that I believe is most important is the selection of the people to help you work on the dream. I would tell my labs that the choice of who they hire as a colleague is much more important than what to work on or what problem to attack. Proper hiring has leverage, enormous leverage if you are good at it.

One of the things that set CSL apart from the other labs at Xerox—and I carried it on at DEC as well—is the way we recruited. Before we elected to interview anyone, we ran their resume or their application or just their history through a rather heavy filter. The field was small enough then; it probably still is, to thoroughly check each candidate. For recruiting new PhDs, which is mostly what we recruited, we would know all or at least some of their professors and some of their fellow graduate students. For someone who is not a new PhD, you have a record of the work they have done and would probably know some of the people they have worked with. You do an enormous amount of homework on these people before you interview them and create a rank-ordered list. You can only interview a limited number of people each year, so you might not interview everyone who's on your list. You start at the top of the list and work your way down. It's very thorough, very systematized.

When a person comes for the interview, the first half day is spent in their giving a talk for a few hours. They give a talk about the work they've been doing or the work that they would like to do or a problem that they think is important. They give this talk before your entire group. There is a tradition and a history of such talks being free flowing and interruptible —even when we give them to ourselves. It's where the "Dealer"[2] concept came from. It's the same spirit.

By the time the candidate has finished a two hour or so presentation, this audience, which is not a typical audience, has put them through some paces and gotten quite a bit of information about the candidate. Then for the next two days, the candidate has one-on-one meetings with as many people in the lab as time will permit.

At the end of this two or three day period, the candidate goes back home and I call a meeting of the lab. We have a thorough discussion from everyone and anyone in the lab who wants to participate about their feelings about this candidate. At the end of that discussion, which might last several hours, I check whether there is a strong sentiment in the group that we really have to have this person come and work with us.

To do this I ask "Who among you would be really disappointed if we did not make this person an offer?" I don't mean just mildly unhappy. I mean *really disappointed*. Now if the show of hands to those kinds of questions was a majority of the people in the room, then we would make this person an offer. And if not, we would not. As a result, you don't make very many offers but when you do, there is a strong commitment on the part of the people who are already there, to the *success* of this new person. Because they've gone on record as saying "Yeah, I really want to work with this person. This person is going to succeed." And that's a commitment to that person's success.

That whole process sets the stage for almost everything else that follows. It builds in a spirit of cooperation. It builds in a spirit about a shared model about what the research objectives are, the objectives of the project. It builds in an automatic, taken-for-granted process for how to work through a problem with your colleagues.

There is widespread agreement that it is important to hire great people. In budgetary terms, labor is the biggest expense in a research organization. But the issue goes beyond budget. The right people can make a research organization work. They have the ideas, and they get excited about them. Since researchers usually stay with a lab for many years, their selection shapes the character and potential of an organization.

Rockefeller University is a graduate and research university in the medical sciences. It is organized in laboratories that function largely independently. The labs vary greatly in size. One has a staff of four; another has about 40 people (some professors, some technicians, some students). As the university's president, Joshua Lederberg found himself engaged in balancing needs of the university as a whole against the more parochial instincts of department heads[3]:

Getting the right people is absolutely central. I have to fight a lot of cronyism and parochialism in hiring and promotion. I think that is always an issue because when a field is not well represented and where an institution may need some invigoration, there isn't any existing constituency. Our faculty include many good citizens and sometimes they will cooperate quite vehemently to try to fix what the institution needs as a whole. However, what usually comes first is their own backyards. How to rebalance authority and collegiality is a very important issue and you don't do it by fiat.

You need a balance between recruiting people, maintaining them and easing them out if the time comes for that. I would say that the importance is about 60 to 30 to 10 in the overall scheme of things. You need to have institutions that are attractive to the kind of recruitments that you want to make. People aren't conscripted into coming. They have to want to come and they have to be the kind of people that you want.

If an institution has been around for awhile, maintenance becomes a little more difficult because the dead wood is very often the liveliest and most powerful laboratory heads. They wield a lot of influence and have lieutenants and in-frastructure that are vitally important to their welfare and the functioning of their own personalities as the lab head. Their interests may not necessarily be congruent with the institution's broad set of goals.

On occasion I had to apply a lot of pressure to block a promotion of an assis-tant professor when such a lab head says that his lab depends on it. I would ask "If we do that for every other lab, what kind of place is this going to be?" Then there might be a glimmer of understanding.

In the absence of replacement, the average age of a group gets a year older every year. To compensate for aging, departments are usually biased toward hiring young researchers.[4] With the exception of research superstars and repeat inventors, seasoned researchers are often consid-ered less likely than fresh recruits to come up with major breakthroughs, because fresh recruits usually arrive with knowledge of the latest methods and technologies. However, at the time of a hiring decision, a young researcher, lacking the longer track record and the more predictable performance of a seasoned researcher, is more difficult to evaluate.

Academic departments reduce risk when hiring young new professors by limiting fresh PhDs to positions as assistant professors, an appoint-ment that is subject to substantial review before promotion to full pro-fessor and typically term-limited to 6 years. Similarly, corporate research laboratories sometimes limit initial positions to fixed terms.

Innovative organizations work best with a distribution of ages in their populations. Organizations made up entirely of new researchers can flounder because they lack the experience to choose optimal research directions. New researchers have less experience in managing projects. When an organization has a sharp divide between older researchers and younger ones, a rift can develop: the older researchers may over-direct the younger ones, and the younger researchers may see little opportunity to rise to leadership positions. Such a social matrix is difficult to sustain. What seems to work best for sustained innovation is an organization with a fairly even distribution of ages.

To paraphrase George Orwell, all researchers are equal but some are more equal than others. Managing an ecology of researchers effectively requires making distinctions. Experienced research managers manage new researchers and seasoned ones differently. Seasoned researchers are expected to take more initiative and to take active leadership roles in projects. Most new researchers need more mentoring to help them to choose projects and make progress. They may take on projects that lack potential or projects that are too difficult. Experienced managers recognize this and hold them to high standards where they learn to articulate their goals and keep up a brisk pace of contributions and publications.

Seeking Stars   As the president of Stanford University, John Hennessy sees hiring and recruiting as critical in shaping the future of a university and determining what kind of effect it can have on society[5]:

I sometimes tell young faculty that what we want to hire at Stanford are people who will swing for the home runs. I think of Babe Ruth standing at the bat. The way a lot of people play the innovation game is that they never swing. Why swing? If you swing you are taking a risk that you will miss the ball and get a strike. So you don't swing. That is a mistake.

Stanford is not a place where we are looking for people who are going to get up and bunt or try to hit singles and make it to first base. Those are not the kind of people who will change a discipline or who will change the world and have a big impact.

Hiring the best people does not always mean hiring people who fit an obvious mold or who will actively work as members of a larger group pursuing a goal. Henry Baird[6]:

I remember the first hiring talk I ever heard was by a young researcher. She was very young and shy and had a faint voice. She had a tendency to cover her mouth with her hands as she spoke so it was almost impossible to follow what she was saying. It was a great effort of will just to hear her. About a third of the way through her talk the hair on the back of my head began to go up and I got very excited. What she was saying revealed a mind like a tank. She was absolutely implacable in her methodology, consumingly systematic, and not in the least bit deflected by fashion. There was no flash. She was just mowing down the ideas one after the other in an astonishing and original fashion. And so I hired her into Bell Labs.

I recognized a quality of mind that I had not seen in a long time, even though all of the appearances suggested otherwise. In fact, even at Bell Labs, for 4 years people thought she was working for me, like my junior assistant. They had no idea that she paid little attention to what I said. The whole 4 years she was off doing her main thing and I would sort of run along side of her and try to connect her with other people. We collaborated a great deal. I'm not saying that we did not collaborate substantially and that I had no influence on her and visa versa. But basically, she was an autonomous intellect and that was the case where I think if I had gone into that talk with strong preconceptions about what a research temperament is like I might not have paid any attention. Or if I had been very controlling, I would not have hired a person who was that independent minded.

As a research manager I may be a bit unusual in that I don't try to steer individual researchers strongly. This may seem extreme, but it is balanced by the fact that for people in industrial research, there is generally a lot of opportunity for collaboration. You have more peers and there's less competition for resources. In academia, very often even within the same department, very few professors collaborate closely. They focus on building their own research programs with their own students and their own funding. It's kind of encapsulated. I think industrial research invites you into a community of peers who are all operating more or less on an equal basis, and are free and willing to collaborate.

Researchers are like the children in Garrison Keillor's Lake Wobegone: "all above average." Henry Baird remarked about his days at Bell Labs that "about two-thirds of the researchers believed that they were in the top third of the performers." This situation may be good for the researchers' satisfaction and productivity because it provides a climate where researchers feel able to take risks and to collaborate with less potential envy. It also reflects a management practice in which there is no public discussion of rankings and no published agreement about evaluation metrics. Although this approach may result in complacency in the ranks, top researchers are generally so self-motivated that they work very hard without prompting from management.

Another reason for this management practice is that research organizations usually have mixed objectives. They are trying simultaneously to deliver results in the short term and to make breakthroughs for the long term. Different parts of the research population tend to be oriented toward different goals and to make different kinds of contributions. Having a single or simple set of rankings and criteria would not respect the diverse ecology of people that make up a research group or the complex choices that a research organization must make in order to thrive.

All organizations need to renew themselves and evolve. A group can grow stale if it focuses too much on its local environment and its special interests. Such parochialism can be a creeping danger. Creating a sense of larger community and purpose in the world requires strong leadership. As Joshua Lederberg put it, members of the Rockefeller faculty tend to look first in their own backyards for hiring. In effect, they are saying "Hire more people just like me." Sometimes the mix of disciplines has to be rebalanced and the purpose of an organization updated. If unproductive people are allowed to stay in an organization for a long time, the demoralizing effects can be substantial and can set a bad example for younger researchers.

Organizations necessarily adopt policies for weeding out low performers. Inevitably, there are some individuals who looked great at the time of their hiring but who, for one reason or another, do not perform well in an organization. A fairly extreme practice is to drop the bottom 10 percent each year. Sensible as such a policy many seem for removing deadwood and improving the quality of an organization, it has unintended consequences. Because new ideas and approaches are inherently untried, breakthrough research requires taking risks. If a research environment evolves so that taking risks seems too dangerous, people shift toward playing it safe. Few breakthroughs result when a culture takes few risks. Thus, harsh pruning policies can have the unwanted effect of reducing both the top end and the bottom end of a productivity scale.

Research organizations should consider whether the skills they look for in hiring will have enduring value. This leads the question of whether fit or quality matters more in hiring. Should a research organization hire individuals who are well suited for a particular project, or should it hire "stars"? As was noted above, researchers tend to stay with a lab for many years. Over so long a period, projects come and go. Since people need to be able to work together in different combinations on a variety of projects, quality matters and momentary fit should not be overemphasized.

Mixing Disciplines
When people from different fields come together, new ideas can arise from the collision of their viewpoints. New ideas occur at the boundaries between disciplines as much as at the leading edges where disciplines are advancing most rapidly. When ideas are bounced around in a group, they mutate and combine; they become sharper and more polished. Thomas Edison—famous for his personal drive—said that he generated his 400 patents with the help of a fourteen-man team.

Innovating at the Edges   The leading edges of a field are where it evolves most rapidly. Innovation happens not only at the leading edges of a field, but often at the intersections where fields come together. As a director at the Institute for the Future, Paul Saffo looks at these intersections in forecasting technology and business futures:

At the institute, it is axiomatic that an advance in a single field never triggers substantial change. Change is triggered by the cross impact of things operating together and not by the impact of a single thing operating alone.

So when you are mapping out technology horizons and making forecasts, you focus on opportunities at the intersections of fields. If you want to innovate, look for the edges. The fastest way to find an innovation is to make a connection across disciplines that everybody else has missed.

Research centers sometimes build cultures that foster collaboration across research disciplines and groups. Without such collaboration, some kinds of projects could never get started. David Fork's "kissing springs" project at PARC, which married gallium arsenide vertical cavity lasers with silicon-based (CMOS) integrated circuits, is an example of such a project.[7] Fork:

The first lithographic spring came out of PARC culture and corridor discussions. Don Smith invented it to be a working element of flat panel displays, but that never happened and the technology was kicking around PARC and languishing to some degree. At about that time Robert Thornton came up with these really dense arrays of vertical cavity lasers, but there was no way to wire them up. You could build the lasers, but you couldn't send drive signals into them to fire them up and emit light. The lasers were fabricated in gallium arsenide, but there was no way to connect them to a drive circuit in CMOS.

The lasers and drive circuits are fabricated separately. The gallium arsenide is used for making the lasers and the silicon chip is for the drive circuits. In materials land, this is often referred to as heterogeneous integration, where you have different electronic devices that need to be combined in some fashion. Silicon cannot emit light and gallium arsenide can't make really good driver electronics; so there is a wiring problem. My idea was to take Don Smith's lithographic springs and scale them down to something really small that you could use to create the wiring. That got a different set of people in different labs excited about this languishing technology.

The original springs that Don had made were maybe sixty microns and the ones that we needed were about four microns. So it was a good order of magnitude shrink, but it didn't require any big change to the technology. It did create an order of magnitude denser interconnect than what had been made available to the industry before. It took Robert Thornton's amazing vertical cavity lasers to require something like that.

People often say that innovation occurs at the edges. This is certainly an example of that. You have some extreme vertical cavity laser technology spurring an interest on an unusual interconnect technology. It pulled in a new group of people to pick up the technology and do other things with it.

We brainstorm all of the time and I think one of the most important things is that we share our ideas and we dedicate at least a small fraction of each new mask design to testing out some of these different ideas. A lot of them are things that you would never tell your boss about until they work. But we keep trying.

In software systems it is not unusual to have a *submarine project* that you reveal when you want to. In material science projects you need to assure access to the fabrication process. It would be expensive and hard to hide bootleg projects[8] if they required their own fabrication budgets. The nice thing about batch fabrication is that the small marginal experiments that you don't popularize ahead of time sit on the same wafer where you are doing your mainstream work. As long as the process flow isn't too different, you're getting a lot of the resource taken care of for free. These experiments are pretty opportunistic in that regard. So it is definitely the sort of thing that you want to encourage researchers to do.

It is common practice for people from EML[9] to sit together at lunch. We have a lot of napkin centric discussions about new ideas. I'll often haul people into my office or I will go and interrupt people with ideas as they strike me. Very often I will wake up at 2 o'clock in the morning and I will jot down ideas until 4:30. Then the next day I will seek out people with the portions of the ideas that seem to be worth passing on. A lot of times you don't find out about people's clever ideas until after they have been tried out on the mask design. There are many scientists that would prefer committing it to a quick fabrication experiment before talking to other people about it. I think that all of those approaches have legitimacy.

It's great when someone shows you something that they have been secretly working on. I think "Oh, my God. You have already made it. Look at that." I think it's really great that we have a process that allows that sort of tinkering to be done without a whole lot of large investment associated with it.

Fork's excitement about progress by others is characteristic of research groups where people work closely together and share results. In these groups, there is a strong sense of community people depend on each other and successes are shared. This is not so typical of groups where people are competitive and sometimes secretive.

Productively Colliding Crafts   How does the joining of multiple disciplines matter in radical innovation? John Seely Brown sees the magic of collaboration as arising from the collision of crafts:

An awful lot of creativity and invention comes from the clashing of crafts, not just the clashing of ideas. That was how it worked in CSL in the early days. You still see it at PARC, especially in the physics labs.

If you talk to film directors, you hear that the collision of ideas happens all the time in the filmmaking. In research, you find much more the creative abrasion and productive collision of crafts. If you look at what is done in solid-state physics and "smart matter," it is the crafts of certain parts of material science coming together with the crafts of computation.

It's not surprising that really fertile areas lie where people have not been before. The breakthroughs often appear in the white space between crafts. This is usually a winning strategy today. But to succeed, you really have to go to the *root* of a problem, what I think of as Pasteur's quadrant. In the process of going to the root, the *pursuit of the problem* pulls multiple crafts together and then these crafts start to collide and in that collision radically new things start to happen. In the early days of the Computer Science Laboratory, you found the collisions of the software people, the algorithms people, and the hardware people all coming together.

The stories in this chapter show how innovation can arise at the edges of disciplines and crafts—that is, in multi-disciplinary projects. Many of the world's leading research centers, including Bell Labs, IBM, PARC, and specialized research institutions at universities, place a high value on having a strong multi-disciplinary mix.

Apparently there is a tradeoff between critical mass and diversity in an organization. The greatest diversity would occur if every member had a different discipline. However, in that case there would not be a critical mass in each discipline. Researchers would collaborate mostly with people in their own fields at other institutions. Most organizations put different weights on different disciplines, focusing on a small number of disciplines but also providing space for a few people of other disciplines whose contributions "spice" the cultural mix.

PARC's research management periodically reviews the inventory of its known[10] projects. One of the properties considered in the accounting of projects is the set of contributing disciplines. For example, in PARC's "content and knowledge" projects, the accounting shows nine main disciplines: psychology, artificial intelligence, computer science, social science, linguistics, physics, economics, interface design, and statistics. In an inventory in 2002, only one project involved just a single discipline. There were eleven two-discipline projects, six three-discipline projects, six four-discipline projects, and three five-discipline projects.

The number of possible interacting edges grows as the number of disciplines increases. Focusing at first on pairs of disciplines, the number of edges between two disciplines goes up more rapidly than the number of disciplines—it goes up as the square. For example, with three disciplines there are three possible edges between two disciplines, but with five disciplines there are already ten possible edges. Depending on how you count, PARC has about 25 disciplines, so there are hundreds of edges for possible pairings. For interactions of three fields in a project, the numbers go up even faster with the number of available fields.

This accounting gives a simple perspective for comparing organizations at different scales and for understanding the real potential of a culture that loves to combine ideas. Very small organizations, such as small startup companies, have very few people, very few disciplines, and thus very few edges between disciplines. They are not set up to generate very many cross-field interactions or ideas. Academic institutions have many people and many disciplines but often fail to exploit the potential disciplinary edges. In very large organizations, such as universities, the distance between individuals can impede the creation of multi-disciplinary projects. Another potent inhibitor can be the effects of impenetrable departmental boundaries. If the culture of an organization erects large barriers to collaboration between fields of research or departments, it will not be effective at exploiting research at the edges.

In summary, the potential for fostering innovation across disciplines depends both on group size and culture. For cross-pollination to be workable, an institution has to be large enough. Furthermore, the culture has to encourage people to cross disciplinary boundaries in working on common problems.

Group Dynamics

By their nature, breakthrough ideas are different from ideas that are ordinary and conventional. People who create breakthroughs become used to working with unusual ideas and practiced in developing and defending them. They sometimes have rough edges and strong personalities. To keep from "blowing apart," organizations develop cultures that encourage people to explore their ideas. Researchers must work together well enough to understand wild ideas, challenge them, and eventually collaborate on them.

Voices of Innovation   The magic of an innovative group comes from the chemistry and complementary skills of its members. Dave Robson, a member of the team that created Smalltalk,[11] saw the dynamics of the interactions between two of the team's members, Alan Kay and Dan Ingalls, as crucial to the team's success[12]:

During the early days of Smalltalk, Alan Kay would arrive at the lab early in the morning, often starting with a shower. He said that he got good ideas in the shower and the water pressure was better than at his house. When Dan Ingalls arrived at the office a little later, the two of them would start out by talking about what they had been doing and thinking about. Alan would talk about his latest thoughts on how Smalltalk or the Dynabook[13] should work, drawing on his vast knowledge of computer science and the arts. Dan would talk about his latest implementation ideas and designs. The practical issues that Dan encountered would guide Alan's thoughts about how to realize his visions. After those morning meetings, the two would often work separately for the rest of the day, working on what they talked about and generating more ideas for the next morning's meeting.

Alan Kay is a great visionary. He talked about a variety of the most important thing to happen with computers 10 or 20 years before they actually happened. When we worked together, he didn't write code, at least not very often. Alan's main products were ideas—presented in his writings and talks. People sometimes had a difficult time understanding what Alan was talking about. His writing was inspiring and poetic, but his ideas were just too far advanced for the time. In conversation, his mind worked extremely fast and it could be hard to keep up with him. In particular, most implementers did not understand how to actually realize what he was talking about.

Dan Ingalls is a great implementer. He has designed and implemented a whole variety of elegant and efficient systems. He also has a great communication bandwidth with Alan. He understood what Alan was talking about and he was convinced that it could be realized. I think that both of them were necessary to create the first generation of Smalltalk. Without Alan, Dan wouldn't have seen that possibility. And without Dan, Alan would not have made it actually run.

One of my favorite stories about Dan is about bitblt.[14] As you know, bitblt is a low-level procedure for moving a rectangle of pixels from one place to another in the bitmap for an image. Everyone at PARC who was writing programs for bitmapped displays knew that writing a general-purpose version of this would be extremely valuable. But they also knew it was extremely difficult because of all the complexities in the underlying bitmap representations—with the word boundaries in memory and so on. Dan said "Yeah, it's hard. But you do it and it's done." He knew that it was an important routine that would be used over and over again—so it was worth the one-time investment to just get it done. When he produced a very fast implementation of bitblt, it made the modern graphical user interface possible—windows, pop-up menus, etc.

I think that in the context of breakthrough inventions, you are often talking about people whose abilities are very exceptional. They are in the 99.99 percentile along some dimension of ability. Such people are obviously very rare. However, many breakthrough inventions require exceptional talent along more than one dimension. And it's even rarer to find someone who has *two* such great abilities. That's why dyads or pairs of people working closely together are so important in breakthrough research. In the case of Smalltalk, Alan was a great visionary and Dan was a superb implementer. Their combined voices were needed for the invention and realization of Smalltalk.

The powerful dynamic between a great visionary and a great practical engineer has driven other great inventions. A very good example of such a dynamic was the interaction between Masaru Ibuka and Nobutoshi Kihara at Sony. Ibuka, one of Sony's founders, was widely known in the company for his ability to anticipate new technologies and to inspire engineers to stretch themselves in trying to meet his goals. Kihara, a brilliant engineer, worked closely with Ibuka to help realize his ideas. In his book *Sony*, John Nathan quotes Kihara:

Ibuka would say that a cassette version would be handier than a reel-to-reel, or that a smaller and lighter machine would be easier to use and easier to sell. He was always thinking aloud about something. And I was always listening. And when we had finished work on a new product, I'd start to think about making that lighter version that he had mentioned. Whenever I had time to spare, I'd experiment on my own and rig something up. I'd have five or six of these prototypes stuffed away under my desk—experimental models. Then one day Ibuka would say "Kihara, a cassette-style portable model would b very handy to have, don't you think?"—and the next morning I'd show him something close to what he was imagining from my secret pile. And he'd be overjoyed. "Kihara, this is just what I wanted!" he'd say. And off we'd go again." (p. 26)

Joint work by Ibuka and Kihara led to more than 700 patents, starting with Sony's first successful product (a tape recorder) and extending through the transistor radio and the Betamax VCR.

Working Together   In teamwork, different members of a team are good at doing different things. Getting a good mix requires more than getting people who can work together well on *one* project. Ron Kaplan believes that groups need to work well on *many* projects over time[15]:

I think NLTT[16] has been a very successful group over 20 years because it has a mix of interests and a mix of people with different strengths. The different interests revolve around a common core of problems. The skills range from linguistics—like Tracy King and Mary Dalrymple to algorithms like John Maxwell and to people in the middle like Dick Crouch and myself and Martin Kay. Individuals like Stefan Reizler bring in a new dimension of expertise about statistical learning.

There have always been very fluid coalitions in NLTT. The thing that I like as a manager is noticing that *these three people* are working together on a problem this week and that next week it's *those three people* working together. Everybody knows that the overall problem is very hard. They know that they only understand a part of it, but they get to work with other people that really understand other parts of it. That's been very exciting for everyone.

Projects come and go, so researchers are not just hired for their fit on a single project. They are hired to join a community where multiple engagements and different combinations of people will work together over time.

Gaining Unfair Advantage   Some of the techniques for gaining "unfair advantage" in research or fostering collaboration are widely understood and have names in a culture. Learning about these practices is a part of the initiation into a creative environment. Mark Stefik describes some of the cultural practices that take place at PARC[17]:

One of the techniques for getting a research advantage is building or getting a "time machine." A time machine is a privileged platform that creates for *today* an environment anticipating what will be widely available in the *future*. You become an early pioneer of the future. You can explore it first, map the territory, and harvest the first results.

When I first came to PARC in 1980, the personal computers that CSL created were time machines. We had these networked, high performance personal workstations for our research that the rest of the world would not have for several years.

When Silicon Graphics was founded, they built advanced 3-D graphics computers for making special effects and for computer-aided design. Stu Card noticed that these computers could be used as time machines since the technology would get much cheaper in a few years and be available for everyone. His group did its visualization research using these pricy workstations. Another example is that when the web first appeared, Stu's group bought a server with a massive set of disk drives. They called it a "web-in-a-box" because it could hold the entire text of the Internet enabling the group to do measurements of web-site connectivity.

As a research strategy, time machines put you out ahead of the pack. There are a lot of computer science laboratories and departments around the world and you want an unfair advantage to get to the leading edge.

Time machines are useful for more than giving an advantage to a research group. A variant of the approach can play a fundamental role in validating a concept for a future product. One of the most difficult challenges for many companies is learning how to take a leadership position in finding a way to "discover the future needs of future customers." How do organizations go about doing this? Often a first step is to engage in radical research related to an important problem—a "pain point"— that is not served by current technologies. The problem should be important enough so that if solved it will have substantial value.

For technologies used directly by people, the process includes user testing and rapid prototyping. In effect, a test bed is created so that a future technology can be experienced. Having users try the technology and carefully observing how well it works for them makes it possible to determine whether the technology would be a "leverage point"—that is, whether it effectively addresses the pain point. Are there additional design issues? Are there unforeseen usability issues or opportunities? The slogan of the time-machine approach is "The best way to predict the future is to invent it."

Understanding Each Other   Cultural behaviors in a laboratory can enhance creativity and collaboration. Bob Taylor:

When there is a stumbling block that a project is facing, there are sometimes various opinions as to how to get around it and so you hash these out. But there still may be disagreement at the end. I invented something called Class 1 and Class 2 disagreements.

A class 1 disagreement is the kind of disagreement that most people have. They just disagree and neither can describe to the other person's satisfaction the other person's point of view. In a class 2 disagreement, each person can describe to the other person's satisfaction the other person's point of view. So at the end of a class 2 disagreement, you might still disagree, but you understand. I will understand to *your* satisfaction *why you believe* what you believe. Those are all elements in facing up to the difficulties in choosing which way to go in a project. Everything lays a foundation for clarity, a huge amount of clarity.

Once a week we had a meeting, called Dealer.[18] It was the only meeting where attendance was required in the lab. Everyone was expected to be there. At these meetings, some member of the lab would be the "dealer" and would stand up and talk about whatever it was that he or she wanted to talk about. The other members of the lab would critique, by interrupting and so on. The reason it's called Dealer is because in a poker game the dealer decides what game you are going to play and what the rules are. In our Dealers, the person standing up that particular day might say "OK, I don't want to be interrupted until I let you know." There might be an argument about that for awhile, but usually there wasn't. Usually the pressure was such that the dealer would say "Sure, I can be interrupted, because I know what I'm talking about. Don't worry about that." So interruption wasn't usually an issue. The point of having a free-flowing meeting was to encourage a discussion that would bring out the lab's creativity and keep everybody on the same page.

Daniel Bobrow has been a member of many research labs at PARC. He has been at CSL, working for Bob Taylor. He has experienced several methods of interaction and argumentation in which people deliberately depersonalize the sides in a discussion in order to foster collaboration and clarity. He reflects on the excitement of working this way across disciplines[19]:

Interdisciplinary collaboration is the most exciting thing that I ever do. It feels like wandering through a space in which I can see some patterns and the other person can see some other patterns. We are looking at the interaction of these patterns and they build something. I am not a musician, but I have this feeling that it is like doing music where the music does not come out of the instrument that you are playing alone. The music comes out of the interactions. For me that human interaction is exciting.

We would get into these discussions. We could play this game where you would switch sides and say "OK. Now trade sides! Make the best argument you can for the other side." I remember going home from those discussions. The ideas were flowing because we were not stuck on a particular idea being our own.

Amplifying the Buzz   Research has a rhythm. People need to spend time alone and time working in a group. Researchers like to work in a group because no individual knows everything. When inventors work on problems that are big enough or new enough, they run into things they don't understand. In principle they could learn about those things and figure them out for themselves, or they could go out into the world and find an expert. But it is a lot faster to talk to an expert down the hall. Mark Stefik:

I often walk into someone's office and ask "Can I borrow your head for a few minutes?" People do this for each other reciprocally at PARC; they act as sounding boards for each other's ideas. Sometimes they have worked on something similar to my problem. They may dig out some variation on the problem or help construct a generalization of it.

This lightweight collaboration happens almost every day. Some days you are working out a small wrinkle on an experiment, and other times you are combining ideas into something big. Feedback speeds up everything and creates buzz.

Buzz is the creative energy of invention. It's the energy of a community of people that are excited because something new has appeared in their midst. When they are all contributing to the same project or at least working in the same space, buzz is a signal of wonderful progress. They are excited and surprised. Everyone is looking at the new "baby."

When people see the new baby, they find out what the baby needs. Metaphorically, they go off and start making clothes for it, or shoes, or something. For a hot piece of buzz you can get lots of minds working in parallel. Everyone has an idea about what the baby will need. This is one of the reasons why people love working at PARC.

When buzz starts to develop, usually people want to share it. When the buzz gets out, more people come around to talk to you about it. There are signals when somebody has buzz to share. They might ask "Want to see a demo?" That's a big clue that they have something new.

Everyone has their social networks. Probably most of the buzz starts circulating in your own research group. But it can jump to other groups. Someone might ask "Who do you know in that group? Who is a good person to talk to?" Researchers have a rough model of who has worked with who and who knows who. This is the magic of a great research center. People are cross-pollinating their ideas. Pollinators are people who spread buzz from group to group. They are buzz amplifiers.

Sometimes the younger researchers keep their buzz to themselves. They love it, however, when older researchers carry their buzz. It enhances their stature. Sometimes people deliberately hide their buzz until they are ready. They may want to work out the next set of ideas themselves, before sharing it. They may be afraid that somebody else will run faster with it, or they may want to make sure that it's polished enough so that it starts out with a good reputation.

Part of the culture at PARC is that most people keep their doors open. If you're at work and you see a lot of people going in and out of an office down the hall, you know that there's some buzz developing. People are dragging other people in to see it. Your buzz antennae go up. You'll probably go check it out. Buzz also spreads at lunch in the cafeteria.

Some people generate more buzz than others. If you are never the source of buzz, then what have you been doing? Some people specialize in helping to bring buzz to reality. They sign up when a project gets going and pitch in. They may not like all the confusion that is part of early buzz, and prefer to work on it when the ideas get more solid.

Reflections

Too much of the wrong kind of management kills off innovation. The cognitive psychologist and computer scientist Allen Newell observed how this can happen for innovative groups[20]:

People think that science is riskier than it actually is. If you take a group of smart people and send them down some path, they generally come back with something interesting. However, one thing that you can do, that prevents them from being successful, is to repeatedly change their direction. Rapid direction changing prevents researchers from sticking with a problem long enough to master the details and make progress.

Management sometimes is unaware of the negative effects of its requests for special "quick" projects and overall changes in direction. Suppose a research manager says "Why don't you talk with company ABC to see if we can solve one of their problems?" and then a few days later says "I'd like you to look into defining a collaborative project with company XYZ." These interchanges and interactions are crucial at some level for open invention. They signal the research organization about what is needed. Switching directions at high frequency, however, is another matter. There can be too many requests to spend time with too many visitors. Such requests are simple to make but can be costly to execute. They take time away from the thinking and innovating budget. If the rate of such requests is not moderated, no research or invention will get done at all.

Making contacts, giving demos, figuring out what (if anything) is appropriate to do, and then reporting back can open up new opportunities, but these opportunities don't come for free. A crash program to evaluate a new opportunity can require several days, disrupting other activities and interfering with coordinated work and deadlines.

A research culture that tolerates too many interruptions—e.g., for administrative meetings, reports, or "dog-and-pony shows" for visitors—breaks the sustained and coordinated concentration needed for inventive insights. When management fails to steer with a steady hand, it saps the innovative vitality of a research organization. The effect of too much interference can show up tangibly as a sharp drop in high-quality inventions produced or papers written, and as a state of lower productivity when researchers don't have enough time to marshal their ideas effectively.

Another interesting quote from Allen Newell comes from a conversation he had with the president of Carnegie Mellon University when he was a University Professor in the School of Computer Science.[21] University Professor is a rank above Professor and one that confers some independence from a particular department. Newell had been approached about being a dean several times, but he thought "deaning" was a terrible idea for him. He was often in demand by the president to be at various decorative or other functions. Newell told him "I will do anything that you want, but not everything that you want." This formulation expressed an exceptional willingness to be at the aid of the university without bargaining about how appealing a particular event was to him. At the same time it established a limitation and predictability for himself as a resource. It cut to the essence of what was important for each man, and it was an attempt to establish a contract based on that.

A central idea of this Newell story is establishing a reasonable budget for meeting these requests. On occasions at PARC when the visitor load for a particular research group gets too onerous, some managers have said "I'll give you three silver bullets for this period of time." In other words, they will support three visitor requests or other "special quick projects" from higher management, but no more. More than that interferes too deeply with other established commitments.

Effective managers learn when to push back on requests, balancing "managing down" with "managing up." Managing down means paying attention to their research group and creating an environment that enables the group to be effective. Managing up means understanding the perceptions and needs of upper management and responding to its goals and requests. A lack of attention to managing up can make an organization seem inadequately responsive and can cause it to miss good opportunities. To pay no attention to managing down is to forget where innovation happens.

Achieving the right balance between managing up and managing down has to do with confidence and inspiration. Success in managing up requires maintaining the confidence of higher management. Success in managing down requires maintaining an environment in which researchers have confidence that they can succeed. Such confidence is undermined when the overhead of task switching becomes too great. When changes in direction are needed, what is called for is *inspiration*. Without inspiration there is constant shuffling around. With inspiration, researchers will build up, sustain, and focus their energies.

10

Obstacles to Radical Innovation

The reasonable man adapts himself to the world; the unreasonable one persists in trying to adapt the world to himself. Therefore all progress depends on the unreasonable man.[1]

—from *Man and Superman* (Maxims for Revolutionaries) by George Bernard Shaw

Kinds of Obstacles

All innovations face obstacles, but some kinds of obstacles are much bigger than others.

The most familiar obstacles are the *innovation* obstacles. These are the "long road" obstacles that all inventions face before they become innovations. They include challenges such as building consensus, garnering resources, mastering the details of engineering, testing, and manufacturing, and developing markets.

Breakthroughs face additional obstacles. Breakthroughs generate buzz and excitement in the world of ideas because they are unexpected and open up new possibilities. A breakthrough idea is outside the scope of conventional thinking, and thus it is difficult, discontinuous, and surprising. The hard part and big obstacle for breakthroughs inventions is *insight*. The challenges of insight in breakthrough inventions are called the *creativity* obstacles.[2] Radical research—the multi-disciplinary approach of following a problem to its root—is a main approach for creating breakthroughs.

The biggest obstacles of all are the ones faced by *radical innovations*. Radical innovations are the ones that lead to widespread change and usually encounter widespread resistance. Disruptive change provokes obstacles. The more radical the changes, the greater the obstacles they provoke. Widespread change takes a long time, typically decades. The big challenges for radical innovations are often economic and social. These challenges for radical innovation are the *change* obstacles.[3]

Some innovations face all three kinds of obstacles. They are difficult to innovate, they challenge conventional thinking, and they require widespread changes before they become practical.

Long Innovation Times

Surmounting obstacles takes time. The time required to surmount the normal obstacles of innovation tends to be predictable. Broadly speaking, development time ranges from a few months in fast-moving industries to 3 or 4 years in industries in which manufacturing and marketing are very complex. Development time is part of the recurring "product-development cycles" of organizations. The amount of time required depends on factors that vary from one industry to the next, including time for design, development, and customer support.

The time required for breakthroughs adds on to the usual time for innovation and is less predictable. Breakthroughs require two major steps: conceptual development and refinement. Estimating the time needed for conceptual development is like the running gag about a carefully prepared meal. It smells good and everyone is hungry. The most impatient party asks "Is it ready yet?" and the cook says "No. It will be ready when it's ready. I'll tell you when it's ready."

Refining a breakthrough invention takes longer than refining an incremental invention because breakthroughs involve techniques that are less familiar and tend to stretch the art of engineering. Incubating and developing a breakthrough technology requires a *sustained* research activity. A representative time to incubate a breakthrough invention through the refinement phase is 2–3 years.

The obstacles to radical innovations are often the least predictable because they arise from resistance to change. Sometimes the resistance arises because the innovation requires a radical shift in how people work—so there can be a very slow adoption curve. Companies with competing old technologies often try to hold on to their existing markets, and mount both marketing and legal challenges to slow down the adoption of disruptive technology. Sometimes the new technology requires developing a set of complementary technologies that provide infrastructure for the new innovation. The time required for the adoption of radical innovation is often measured in multiple decades.

An important consequence of the time delays for breakthrough and radical innovations is that the sponsors for these innovations sometimes need to wait a long time before they can realize commensurate economic benefits from their investments. In a corporate world where shareholders pay attention on a quarterly cycle to return on investment, the need for quick results often gets in the way of the patience needed for breakthroughs. When companies lack the resources to refine breakthrough inventions or the patience to shepherd radical innovations through early markets, the main benefits of the innovation may be lost or captured by others.

The benefits of breakthroughs and radical innovations are widely shared in society. Such innovations make the future potentially much better than today. However, in the absence of unusual and favorable circumstances, the innovations that contribute most to our sense of progress are the most difficult to pursue.

Stories

Antibodies to Innovation

The Internet was made possible by *packet switching*.[4] When someone sends email, reads a web page, or downloads music, computers break up the information into small packets. These packets are exchanged over communication lines by the computer nodes of the network, from one node to the next, until they arrive at their destination; they are then reassembled into the complete message.

Packet switching is now the standard method of sending digital data.[5] Robert Kahn was one of the inventors of the communication protocols for it. Although the industry now takes packet switching and many other technologies for granted, they faced huge resistance before they caught on. Kahn[6]:

There are a lot of inventions that don't make it into the market simply because there are forces that don't allow it. For example, packet switching did not get adopted in the late 1960s or early 1970s when it first was worked on. When you look back now, 30 years later, virtually every communication system that has been built in the world since then has been built using packet switching. It is used not only for computer communications but for managing internal resources as well.

Packet switching didn't get adopted right away because it was too challenging or too threatening or so counter culture that nobody was basically able to make a decision to use the technology. I think this is a generic problem with most innovation: The more serious the innovation the more likely that there will be the resistance to its acceptance fighting a push for its acceptance. It's as if the system has antibodies.

A lot of the things I've been interested in have been about *infrastructure*. Changes to infrastructure tend to upset whole areas of business. Consider what happens if you put a railroad or a light rail into an area. Right away the land around it becomes extremely valuable as compared to other land just by virtue of access to the infrastructure. New infrastructure tends to induce broad changes and often causes many kinds of push back.

There are many other examples of technologies that were difficult to adopt at first. Why didn't time-sharing computing get adopted right off the bat? When you look at the big computer companies in the early days, they were places like GE, Honeywell, IBM, and Burroughs. Time-sharing systems were very threatening to IBM back in the early days, because, instead of selling lots of different machines they would end up selling one that could be shared. Time-sharing had a potential to undercut their business. Yet eventually the economics of the whole field tended to force this as a long-term direction. I am sure you could think of lots of examples where a new technology created competition for some entrenched part of a current market and was resisted.

Comfortable Ruts as Obstacles

Bob Taylor was manager of the Computer Science Laboratory at Xerox PARC during the 1970s and the early 1980s, a period of enormous creativity and inventive output. During Taylor's tenure, CSL created the personal computer. The system hardware included the bitmapped display, the mouse, the Ethernet, the file server, and the laser printer. The system software included the graphical user interface with menus, modern text editors, drawing tools, electronic mail, and visual networked shared-reality games. At our request, Taylor reflected on obstacles that make it difficult to adopt inventions[7]:

Many of the obstacles that we faced were cultural. For example we created this Boca Raton exhibit to promote the personal computers to Xerox. It was a several day meeting in Boca Raton, Florida, of the top 250 corporate managers at Xerox and their wives. The last day of this four day extravaganza was devoted to the future of Xerox. In an auditorium with a huge display and lots of fan fare to an audience of about 500 people, we demonstrated the inventions that we had created at PARC. We demonstrated the Alto, the Ethernet, laser printing, Bravo,[8] desktop publishing, and SmallTalk.[9]

It was quite a show. We had professional producers and scriptwriters and advisors helping us put the story together in a Hollywood type spirit. In fact we rehearsed the show on a Hollywood sound stage in Southern California. In the afternoon after the show, we had demonstrations off to the side. We set up demonstration areas so that husbands and wives could walk to an Alto, sit down and try things, and go over to a laser printer to see the results. It was a hands-on adventure.

This Boca Raton event took place in 1976 or 1977. It was then that I began to see the writing on the wall. This technology was not going to be an easy for Xerox to understand. What started me thinking along these lines was as people walked through the exhibits in the afternoon, it was the wives who sat down and tried things. The husbands stood back and watched. The wives thought the technology was wonderful and the husbands were silent.

I came to realize that the wives had jobs in offices dealing with the document preparation steps that our systems addressed. The husbands were high-level managers and thought it was beneath them to sit at a keyboard and type. The wives understood the differences between how people did things in offices in those times and what the Alto and the system permitted people to do. They were really enthusiastic about it. The husbands were absolutely blasé. The husbands were also working for a copier company, not a computer company. Copying was still making them a lot of money and they were very successful. They said "Why should we be interested in this other stuff?"

The managers lacked vision. They also lacked imagination and courage. They were in a comfortable rut and they wanted to stay there. That situation would describe Xerox and a lot of other companies. Once companies are successful, they find it next to impossible to embark on a slightly different direction to guarantee their success down the road. It happened to Xerox; it happened to IBM; it happened to DEC.

The only thing that I know of that could have kept it from happening, and then there is no guarantee, is to set up a place like PARC as a skunk works. By that I mean, a place that is insulated from the rest of the company, for awhile anyway, so its management can not muck with it. It has completely free rein; it has a reasonable size budget; and it can do what it wants to do. This is the way that some very important new technology airplanes were developed by Lockheed in Georgia. They set up a skunk works. They took a group of aero engineers and put them off in the skunk works protectively. The U2 was designed in the skunk works. Later the [SR-71] superseded the U2.

There was a plan in the sense that said "Hey look. You are going to build this special airplane that is not like any other airplane that has ever been built. So you throw out the window a lot of preconceived notions. You try to build something that has these properties, but that may require you to develop new technologies."

This is called "disruptive technology." If you want to be successful, or if you are successful now, and you want to continue it, then you had better be the one that develops the disruptive technology. If someone else does it, you're dead. That is what happened to IBM and Xerox on the personal computer.

From Taylor's perspective and with hindsight, Xerox managers lacked vision. Of course, many of the Xerox managers were very hardworking and had plenty of vision in the areas where they were focused. They were busy growing the copier business and their attention and creativity was focused there. They did not have time and did not see the urgency or value of turning their attention from a rapidly growing and profitable business to develop a new business around a risky and radical invention.

Xerox's inability to successfully deploy the personal computer has been the subject of several books. There was no entrenched interest blocking the deployment of personal computers—they didn't displace any systems that Xerox cared about. However, they were viewed as a distraction from the very profitable copier business, in which Xerox had a near monopoly as a result of its patent portfolio. As Taylor put it, Xerox executives "were in a comfortable rut and they wanted to stay there." It can also be said that Xerox never figured out how to market the computers. When Xerox sold its first personal computers, it was nothing like the way that other companies sold PCs years later. A customer could not just buy one personal computer. They had to buy many of them all at once, together with a laser printer and a file server.[10] In short, the purchaser essentially had to equip an entire workgroup with the new technology, wire the building for an Ethernet,[11] and train people in a new way of working. Very few people in big corporations would write the check for a radical adventure with the early Xerox computers.

Xerox had faced a similar conundrum when it first tried to sell its copiers. As with personal computers, the early copiers were so expensive that most potential customers were reluctant to shell out the cash required for new copier. They believed that they wouldn't make enough copies to justify the expense. Customer attitudes were shaped by their experience making copies using carbon paper and mimeograph machines, where the number of copies made was quite limited.

Xerox believed that customers would find many new reasons to make copies once they had a convenient copier in the office and experienced the high quality of the copies. They developed a plan where the cost depended on how many copies people made. The auditron devices—the hand-held counters that used to be part of the ritual of using office copiers—were an invention of marketing. Each customer department had its own auditron. Every month the auditrons indicated the number of copies that had been made and Xerox billed companies accordingly—by the copy rather than for the copier itself. It turned out that people found many creative and useful reasons to make copies. Auditrons are credited with making the explosive growth of Xerox possible. They enabled people to take a low risk in using a new technology where they were not sure how useful it would be. In summary, Xerox was *capable* of break-through ideas in marketing and demonstrated this in its copier business. Like most companies in a similar situation, Xerox became trapped in the innovator's dilemma.[12] Xerox found it difficult to take attention away from its rapidly growing business to successfully build a second business around another radical invention.

The Good versus the Best

When Ted Selker worked for IBM, several of his inventions had to do with laptop computers. In this story, Selker focused his attention on their power supplies[13]:

One time when I was showing an invention about a pointing device to a high-level executive at IBM, he said "But what about this problem with the power supply? It has all the same attributes as the mouse. It has this tangly long cord, and it's got this big object. How do I get rid of that?"

I had been thinking about power supplies and had an idea to simply snip off the part of the power curve that is below 20 volts and dump it into the battery of a notebook computer as though the battery was a capacitor. I told him this idea—which was to make something that got rid of the capacitors and inductors, each of which was a third of the power supply.

The idea was to use the capacity of the battery to eliminate these parts. From a high-level view, transformers store energy for part of a cycle and then dump it out later. A battery can be charged by pulses as efficiently as by continuous power, which was my hypothesis. The other thing that I learned was that my idea got rid of most of the bulk of the power supply. I hired people to implement it. One of the interesting things was that along the way of implementing it, everyone including the power engineers didn't believe it was a valuable or useful thing to do.

But I never gave up on it, and continued trying at the same time, solving a lot of other problems. For example, one of the important problems about a power supply is that you want to separate the high voltage. If a voltage is above $47\frac{1}{2}$ volts, it has to be physically separated from the power user in modern electronics by 4 millimeters. Conventionally this is done by having the transformer coils separate from each other. I had an idea to split the battery into two batteries and have a relay whereby we are charging one battery and taking power off of the other. That way the battery that is supplying power to the notebook is never physically connected to the 110 volts. I was very proud of this.

I like to think of a *hundred* different ideas if I can. So usually when I'm solving a problem I'm building two or three different examples of solutions, hopefully coming from different points of view. I worked on several parallel projects for improving laptop power supplies. These included a transformer that was integrated with the power cord but very inefficient, and a very efficient transformer with tiny cores and long coils that emitted significant electromagnetic radiation. I also found some simple ways to improve power supplies by making small changes in their industrial design.

We found three ways to make the power supply smaller by looking at the industrial design. I saw that the strain relief made them 3/4 of an inch longer, I saw that the way the cord plugged into it made the whole thing volumetrically about 20 percent larger, I saw that the EMC, electromagnetic radiation core on the cord added something that was twice the size of the handle that you held that plugged the cord in, so I made the electromagnetic radiation thing *be* the handle that you plug in the cable connector. I figured out how to put the strain reliefs inside. And then I invented a plug that will plug in anywhere in the world and select automagically the right prongs by the pressure into the wall.

At this point I made the mistake of not solving *any* of the problems completely enough for them to buy into them. The product manager said "We can implement this new cord you designed with our industrial designers. The other technological solutions are harder. They would cause us to make a development team that creates power supplies that are non-standard to the rest of the industry. We *like* buying a Panasonic power supply for 10 or 15 bucks." So in the end, the radical innovations were difficult to take forward because the development organization had its own constraints and wanted to do the simplest thing.

In the end, they picked only the things that changed the industrial design and cost them nothing but shape. Still, these changes did make an improvement in the product and satisfied a bunch of executives and everyone else. To this day I do not know whether I was better off showing them all of those possibilities, or whether I should have parceled them out.

Selker's story illustrates the approach of following a problem to its root. The hard problem was brought to him from a product division. His pursuit of several competing solutions to the problem involved bring in several disciplines. This included engineers and mathematicians for finding ways to get rid of transformers. It also included industrial design expertise for the simpler solutions that made the conventional design more compact. By pursuing several approaches at once, Selker's team was able to create a range of options for the product team. Even though the most radical approaches were ultimately not chosen, the team explored them far enough to illuminate both the real obstacles and the costs and values of surmounting them.

Embracing the Unconventional

David Goldberg is a mathematician and an inventor at PARC. Goldberg is known for his invention of Unistrokes, an early version of a simplified writing alphabet much like Graffiti for the Palm Pilot.[14] The approach is called Unistrokes because each letter is written using a single stroke without lifting the pen.

Goldberg started working on Unistrokes in the context of the ubiquitous computing (Ubicomp) project at PARC. Ubicomp was a project where researchers experimented with populating the work environment with interacting computers ranging from tiny tabs to wall-size liveboards.[15] Researchers on the Ubicomp project used computers of all sizes in their daily research lives. They used liveboards in meeting rooms across PARC and wore "tabs" that signaled to the system where they were in the building. They used palm-size computers to read and respond to email as they walked around or sat in meetings, and developed pen-based tablet computers for taking notes in meetings.

When we asked Goldberg how he got started on Unistrokes, he described his idea of "finding a hole" in an ongoing project[16]:

There are several ways that I like working. One of them is to look at a project and say "Oh, look. They've left a hole." I look for something to do that fits into the hole.

The Ubicomp project was underway and I asked someone "If you have a small device, you can't have a keyboard. What are you supposed to do?" They said "I don't know." So that was step one. I found a hole. That is fun because you can have your own problem but you are also part of a bigger project. Finding a hole is a strategy that has worked for me previously.

Another thing that often works for me is to repeat work that was done earlier in order to understand an area. I knew that other people had worked on handwriting recognition. I wasn't trying to be original. I found some papers on handwriting recognition and wrote a program based on algorithms from the literature. That's a good way to come up to speed. Although you're duplicating what's already been done, it's easier to go forward when you have some idea of what's already been done.

After struggling with handwriting recognition for awhile, I thought "this is a dumb way to do this. It's *never* going to work well." Maybe I was lazy and did not want to spend the next 5 years of my life writing a good handwriting recognizer. Maybe that's what was really going on. But I thought "No, I have a good idea here and all these other people are wrong." That's very exciting, realizing that everyone else is going in the wrong direction. At least you think they are.

Goldberg's idea of creating a new alphabet was criticized by almost everyone he talked to. Many people felt that computers should change to make things easier for people. It seemed wrong to make people change how they write in order to make it easier for the computer to recognize the writing.

Mark Stefik worked with Goldberg on a predecessor to Unistrokes called "heads-up writing." The Heads-up Writing project was born out of the observation that taking notes on laptops during meetings was distracting. It is distracting to the meeting in part because of the clicking sounds of keyboards.[17] Pen input was a potential alternative, but visual attention is still required to keep the pen on a line and to reposition the pen at the end of a line. The heads-up idea was to write letters one on top of the next in a small space[18] without moving the hand to the right. A technical challenge was to find a reliable way to determine where one letter ends and another begins. For example, the system must distinguish 'l-' from 't'. When Goldberg suggested sidestepping the character identification problem by creating a new alphabet, Stefik moved on to other projects. He thought that learning a new alphabet would force users to make too many changes in how they work:

The Heads-Up project helped me to think of the single-stroke idea. As the project got going, people began to react to it. Some people would say "Oh, how intriguing." And others would say "Well, that's ridiculous. No one's going to want to do that." For example, I remember when Dave Patterson[19] visited CSL. Patterson thought that the whole thing was ridiculous. He said "Speech recognition—that's what you should be doing." But other people didn't think speech recognition was a good idea.

Almost everyone who tried it said "I thought it would be hard to learn a new alphabet, but it's actually easy." So that was great. It got me over this hump from all the naysayers.

When the work was all done, I wrote a paper and sent in an invention proposal (IP).[20] I had done other IPs. I thought this was one of my *better* ideas. The report came back from the TAP panel[21] with a 2 rating [on a scale of 5]. This is about the worst rating you can get. You never get a 2. A 3 is normally the lowest you ever get. I had never even *heard* of anyone getting a 2. And then there were comments, which I am sure were sanitized by Chuck Hebel. They said something like "This is a really stupid idea. No one is going to learn this." And then it said something like—it didn't use these words, but the gist of it was "Since obviously you can't think of any decent research projects, here are some things that would be more valuable to work on."

I'm very proud of the TAP panel report. Years later I hung it outside my door.

I told my boss, Mark Weiser,[22] about the TAP rating. I said "They didn't want to patent it. What should I do?" And Weiser said "OK. Then just publish it."

So I wrote a paper for CHI[23] and someone suggested that I make a video, which I did. I gave the video to Weiser. Every time I saw Weiser I'd ask "Have you watched the video yet?" And he'd say "No. Not yet."

One day I came in and Weiser came over and said "Wow. I saw that video. Have we patented this?" Then I reminded him that the TAP panel hated it and he had told me not to pursue it. Weiser got on the phone to Tom Webster[24] and got him to write a patent application.

Later I presented a paper about Unistrokes at CHI. I rarely go to conferences. I had never met Donald Norman,[25] a user-interface big shot who used to be at the University of California at San Diego. After you gave your talk at CHI, there was a discussant. I exchanged e-mail with Norman and knew he would be the discussant. I knew he had written a book, but I had never met him.

During my session there was also another paper by a student of Bill Buxton,[26] which had a one-handed keyboard. And I can't remember the third one. So now it's time for Don Norman to discuss them. He launches this long tirade about how terrible all the papers were. I wouldn't have minded that if I had a chance to rebut. I'm happy to have a discussion. But the way it was set up he got the final word and he was really nasty. In some sense it was good. He was so nasty that I didn't take it seriously.

The Buxton student was so upset that the moderator actually let him rebut. Half the things that Norman said weren't even correct, just lambasting his work. But that was just kind of a funny little incident, which Tom Moran often jokes about because he was there at the talk.

When I gave talks about Unistrokes, I found that people who had worked on the handwriting recognition problem were almost always much more impressed than other people because they immediately thought "Oh, yes. This solves the problem I've been working on." Whereas people who didn't know anything about handwriting recognition tended to think that a new alphabet was not a good idea.

At that time, if you had told me that there would be millions of people using Unistrokes and that this would eventually be worth millions of dollars, I would never have believed it. I thought it was a very good idea, but I have a *lot* of good ideas. Most of them never go anywhere. I never guessed that this one would take off.

Goldberg's story shows how conventional thinking can get in the way. The very idea of having people learn a new alphabet was absurd not only to his colleagues at PARC, but also to people at the CHI conference when he presented a paper. Goldberg did not come to the idea of changing the alphabet to be cute or different. He got there by exploring the alternatives for building a recognition system for handwriting. Even when he had the solution in hand, it was hard for people to accept the idea that it was reasonable.

It is noteworthy that all the resistance David Goldberg faced with his invention took the form of negative peer review. When Stefik abandoned the heads-up display project, he was giving a sort of peer review known as "voting with your feet." When visitors criticized the project, when the PARC invention panel rejected the idea, and when Don Norman criticized the research at the CHI conference, each of these events was an instance of negative peer review.

What sustained Goldberg and gave him confidence to keep going? Perhaps the main reason was that when users tried Unistrokes they reported that it was not difficult at all to learn. Similarly today, people using the Palm Pilot products typically discover that they can begin using Graffiti right away. Contrary to the expectations of Goldberg's peers, learning a writing style as an adult is not nearly as difficult as learning the big ideas of an alphabet and reading as a child.

This experience also gives some insight into the value of having a mixture of methodologies and professions in a research center. Although the concept of peer review is central in organizations grounded in science, it is less central in engineering and design. When Goldberg evaluated his approach by having users try it, he was following the methods of engineering and design. Those fields place less confidence in peer review as a method for evaluation. Peer review assumes that we can evaluate something by thinking about it. It is therefore subject to the biases of preconception and hidden assumptions and other limitations of the imagination. User studies, on the other hand, measure the usability of a thing itself. Users don't need to imagine the device—they can experience it directly. Consequently, user studies are inherently freer of that kind of bias.

Doing the Impossible

Digital property rights is a technology for regulating the distribution of digital works, such as digital music, or books, or newspapers. Mark Stefik believed that such technology would enable authors, artists, and other creators to turn their works loose into a marketplace and make a living in this way. As with other radical ideas, deploying the technology has implications for the structure of the publishing industry, as well as for the legal and economic infrastructure that supports it.[27] Stefik:

The first obstacles for digital property rights appeared immediately inside PARC. I had filed an invention proposal with pretty ambitious expectations. The conventional wisdom at PARC and elsewhere was that the digital copyrights problem was unsolvable because digital bits could always be copied. Colleagues listened but they believed that there had to be a flaw in my idea. There wasn't as much familiarity with cryptography ideas in those days and people figured that I was an Artificial Intelligence guy working outside my area of expertise, which was true. What did I know about cryptography or security?

John Seely Brown was director of Xerox PARC at that time. He knew that I was reasonably smart, heard the buzz, and felt that the idea could be worth a lot, but only if it worked. He asked Ralph Merkle to look into it. Ralph was one of the inventors of public-key encryption systems and is one of the best crypto guys in the world. He read over the invention proposals and thought about it.

Ralph has this deep basso profundo voice and sometimes takes awhile to get to the point. Or maybe I was impatient. Anyway, he came by to talk with me about it a few days later and started telling me about some related earlier inventions of his. I remember interrupting his story and he said "Well, in a word, it works!" That changed everything. People said "Wow!" I was invited to various labs around PARC to talk about it.

That's when digital rights went from being impossible to simply being "immoral." This was the next big obstacle. People in the computer labs at PARC grew up as part of the academic computer science culture and hacker culture. Copying programs and sharing data had always been a key both to building scientific reputations and advancing the art. Traditionally, PARC had built its own computers and written its own software. It was transitioning to using commercial workstations and personal computers and was getting used to actually *buying* software, but the logic of that hadn't really seeped in yet.

People kept coming up with objections. "What if I have to make a backup copy of something I bought? I ought to be able to protect against disk failure. But if I can make a backup copy, I can make any number of copies, so this will never work." These objections were gifts of insight. I used them to refine the invention.

Basically, the social shift was too big. The PARC researchers didn't like a world where software wasn't free. Of course they were living in that world already, but they didn't understand yet how the rules were changing. In general, though, many people thought that the ideas were intriguing. Most of the scenarios of use that I was working on had to do with buying or loaning books, or movie tickets, or music. Napster—the music and file sharing service—hadn't come and gone yet. I was busy explaining how you could do e-commerce. Interest in e-commerce was heating up as the dotcom boom started.

My best contacts to the publishing industry were to the book publishers. We created a team of conspirators inside Xerox trying to get the technology out. Prasad Ram led an advanced development team. Bettie Steiger led the development of business planning. Prasad and I made some trips to New York City under the auspices of the American Association of Publishers. Their technical leader, Carol Risher, was always trying to help the industry find its way into digital technology. She put us in touch with early adapters and opinion leaders.

The publishers were interested in digital property rights, but there were an amazing number of obstacles. I think that the most serious one was that digital books require a reading platform. There wasn't a usable and affordable book reading platform anywhere in sight. Absence of a ready platform for electronic books made the conversations very short. People felt that nobody wanted to read books on their PCs. If there was no cheap book reading platform, why were we talking about digital distribution and digital rights?

I discuss all of this in my book *The Internet Edge*. A book reading platform requires a long lasting battery, a clear and cheap display, as well as other things.

The same platform would work for movies, newspapers and music too. However, when you follow the technology curves that describe when these technologies will be ready for mass use, the prospects were a few years out.

Without an electronic book reading platform, the main business opportunity was on-demand printing of low volume books. You could go to a bookstore, order a book, and they would print it on the spot—and the publisher would be reliably paid. I wanted to explore opportunities in music distribution and also secure distribution of internal corporate documents, but Xerox wasn't interested.

Inside Xerox there were also many business obstacles. Xerox had a Corporate Innovation Council that was responsible for evaluating and funding proposals to create new businesses. We were never able to get more than seed money from this organization to launch the operation.

There were also major legal obstacles. Basically, the legal system wasn't and still isn't quite ready for digital distribution and digital rights. Copyright law tends to lag technological change, and it has a hodgepodge of rules and exceptions. There were all kinds of special cases about broadcast and things that had shaped different industries differently—such as the differences in the rules between delivering music using a conventional FM radio broadcast and delivering it on cable. Digital networks made the old distinctions unworkable.

Another issue was that fair use in copyright law needed reworking. Fair use is the idea of being able to quote without cost small portions of a work for basically non-commercial purposes. Digital property rights technology shifts regulation from copyright law to contract law, where there is no established notion of fair use. The teachers, composers, and writers who wanted to exercise fair use were typically not capable of covering the losses if unprotected copies of a work got away. Imagine Professor Jones at a local community college is using a digital copy of *Star Wars* in a movie appreciation class. He makes some technical blunder on his PC, and an unprotected copy of the movie gets out to the Internet and spreads around the world. Potentially hundreds of millions of dollars in revenue are at risk. Professor Jones is not prepared to cover the losses.

The legal issues also included handling import and export controls for anything on the net—since it crosses international boundaries so easily. I published a paper[28] about this with an attorney, Alex Silverman, making some broad proposals about the kinds of changes in the law that would make sense in an approach based on contract law.

Since then, many people have gotten involved in the legal tangles. The controversial Digital Millennium Copyright Act has been enacted to address some of the issues—but there will be some challenges in the courts and it will take years to clean up this mess. A lot of people have a stake in the current complexities of the law and how it turns out. This includes people in different industries—movies, music, broadcast, software, and books to name a few—and also attorneys and interest groups in different countries around the world.

The world is engaged in getting ready for digital rights. For example, the music CD business is changing now that kids can burn their own CDs. The amazingly broken security model for DVDs came about because the people pushing the standards would not listen to the computer people who had the expertise. I remember giving a talk at a couple of the industry symposiums on DVDs before they came out, urging them to use stronger security and also to consider incorporating digital rights management right from the start. Folks from Intel, Macro-Vision, and a few other companies understood this, but the entertainment people were not too interested and the component people were too far along to change their plans. The security of the DVD system depended on the secrecy of a single number, which was embedded in every device. That made it very vulnerable. Al Gore was right that the digital copyright problem needed addressing, but the hardest problems were not technological.

After we filed the patent applications for the digital rights technology, I wrote a paper called "Letting Loose the Light"[29] which sketched out what I thought it would take to create a vibrant market for digital works on the net. I proposed creating an organization called the Digital Property Trust that would involve the stakeholders from the connected industries. I still think that is the shortest way to adopt digital rights management. If the relevant companies start a patent war and don't pull together, adopting digital rights management could take a long time.[30]

The story of the invention of digital rights management illustrates the approach of pursuing a problem to its root. The project started with the identification of a challenge problem of potentially great value to "The Document Company."[31] In the following to the root approach for breakthrough research, the enlisting of disciplines to penetrate obstacles is a key part of the approach.

After the initial insight, Stefik encountered obstacles in the detailed design of communications protocols. Computer-security researchers had long recognized several kinds of attacks on communications, such as spoofing and the "man in the middle." To bring in a deeper knowledge of cryptography and the design of communications protocols, Stefik sought advice from others at PARC.

Additional researchers became involved to explore designs for trusted portable hardware for documents and other digital works.[32] The very first set of insights involved protecting digital works but did not include a financial model. Models for financial transactions needed to be developed. The project expanded to include a social scientist and others to work on scenarios of use for digital publishing, and in this way to establish some of the use parameters for financial transactions. This work uncovered a range of financial models that the rights language would need to cover. Later in the process it became clear that there were many deep obstacles in the legal system to digital rights. The concepts of fair use that were controversial in when information can be used without fee were grounded in copyright law. However, the trusted systems were fundamentally based on contracts and related to contract law. Contract law has no provisions for fair use, and it was clear that there would be a battle among the stakeholders if this wasn't sorted through. Ultimately the solution would have to address risk management, perhaps with the participation of the insurance industry. An attorney joined the project to help understand the deeper issues and think through the alternatives. A follow-on project at PARC explored the concept of "trusted printing," which brought in additional disciplines involved in the design of embedded data for watermarking and exotic printable inks and papers to make documents difficult to copy. Members of the project team interacted with bankers and publishers to deepen their understanding of what form a solution should take.

A radical research approach was taken. As obstacles were encountered, the project added disciplines and people to understand them and try to crack them. The added disciplines and competencies included cryptography, protocol design, electrical engineering, user interfaces, social science, law, and materials design as well as business people with expertise in publishing. In this way, a multi-disciplinary team created a fast-track process for understanding digital rights and creating a potent patent portfolio for Xerox that anticipated the needs of the market by several years.

As the 21st century opened, many companies and publishers became increasingly interested in digital rights management. The topic has been covered often in cover stories in mainstream computer magazines and in legal and business journals.

In April 2000, Xerox and Microsoft formed ContentGuard, a joint venture owning the patents and technology.[33] In 2003 the industry began to consolidate, with Sony and Philips purchasing InterTrust and licensing technology from ContentGuard. The entertainment industry is still struggling with the downloading of videos and music. A new round of exploration by Amazon and others to bring the contents of books online was also surfacing.

Radical Research and Federal Funding

The U.S. government plays a major role in innovation by selecting and funding of basic and applied research. This funding is managed by the National Science Foundation (NSF), the National Institutes of Health

(NIH), the National Institute of Standards and Technology (NIST), the Defense Advanced Research Projects Agency (DARPA), and other bodies.[34]

Like all participants in the innovation ecology, federal research granting organizations ask "What is possible?" and "What is needed?" In broad terms, their missions are to improve the public good by advancing knowledge, to contribute to the growth of the economy, and to ensure national defense. They allocate their research budgets across multiple fields of study and across both basic and applied research. Focusing on research rather than on products, they try to anticipate which research investments will most benefit the public good.

Hazards of Evaluation Metrics   In funding research, it isn't always possible to determine in advance how discoveries in various fields will unfold or which research will have the greatest benefit. When expectations are high for an area of research but progress seems slow, the desire may arise to bring more focus and direction by managing goals and objectives more closely.

When sponsoring organizations want to accelerate the pace of progress in an area of applied research, they can sometimes try more direct methods to steer it. However, steering research too directly can be counterproductive. It is not typical to use results-oriented competitions to set a direction and to determine the awarding of funds for research. In the late 1980s and the early 1990s, such an approach was used to drive research in natural-language technology in the Message Understanding Conference (MUC) competitions. Jerry Hobbs has seen that some ways of attempting to measure and steer progress in research end up killing off the diversity of ideas needed for breakthroughs[35]:

In the 1980s much of the research in natural-language technology kept doing the same thing over and over again. People would create yet another grammar with sub-categorization constraints. There were few practical applications. DARPA thought that researchers were just spinning their wheels and decided to try to drive the field with focused competitions. One of the things that the MUC competitions did was to push natural-language technology closer to real applications.

DARPA was trying to figure out how to evaluate and judge progress. [In 1991] it was a struggle for natural-language systems to deal with understanding a whole paragraph. When you scale systems up from recognizing words to understanding paragraphs, the systems need to use a surprising amount of knowledge. Every piece of knowledge connects to a lot of other pieces and you almost have to know about "half of the world." The curve of how much systems need to know is very steep. We hoped it would flatten out. We were on the steep part of the curve, which meant that we were doing a lot of work encoding world knowledge and not getting many results.

We were proposing to analyze about a dozen texts. We wanted to do them deeply and to develop general principles. The plan was to generalize the analytic machinery and to speed it up later. The funder's response was "Your plan sounds good, but we want to see a thousand texts instead of a dozen." Of course, at the scale of a research grant to analyze a thousand texts precludes doing any of them deeply. That would require far too much world knowledge. DARPA's request forced all the research groups to use superficial methods rather than attempting any deep methods. That is what happened throughout the 1990s. There was a very strong push toward superficial methods.

This approach led to a 5 or 6 year detour for science. For example, one researcher was very interested in verb-phrase ellipsis, sentences like "John called his mother and so did Bill."[36] These sentences occur in natural discourse and this is an example of the kinds of phenomena that linguists want to model. However, even if you did a perfect treatment of verb phrase ellipsis, it would probably only raise your score on the competitions by 1 percent. Tackling such theoretical problems is difficult, and it is hard to do that kind of research in this kind of funding program.

The problem with shallow methods—the sort that the funding model encouraged—is that they top out. You can't get to higher levels of performance without encoding the deep knowledge. Why does it top out? That is a good question. One possibility is that the set of problems are in a Zipf distribution, so that the first ten problems that you deal with are very frequent and they get you 40 percent of the way there. Then solving the next ten problems gets you 60 percent of the way there. At that point there are so few instances of each problem so that anything you improve has little effect on your scores.

We had this one project where our job was to spend a year seeing how far we could push the technology. We were working on that and NYU was working on that and in a month you can go from 0 to 55 percent and in another year you can get it up to 65 percent. The typical experience is that you would look over your mistakes in the training set and find a couple examples here where the system didn't handle plurals right or something. You fix that, but then you run the system on the competition set and there is no improvement in the score at all because those linguistic phenomena do not happen to occur in the data. This is a common phenomenon once you reach a certain level.

Two of the key measures used in the competitions were precision and recall.[37] In 1991, the best of the systems were around 50 percent in precision and recall. Then everyone copied the methods of the winning systems for the competitions in the next year. In 1992 they had pushed performance up to 60 percent. In 1993 the best of the systems were still at about 60 percent, but there were more systems in that range. In 1995 all the systems were in a tight little cluster a little below 60 percent recall and a little above 60 percent precision. The main difference between that and 1998 is that in the latter year, fewer sites entered the competition. The effect of the evaluations was to drive the systems to a local maximum. The technology plateaued.

This highly directed approach to funding reduced the diversity of approaches and risk taking by the researchers. Any successful innovation that one participant had developed and displayed in an evaluation would be copied by the other participants by the next evaluation. In fact, the other participants would have to copy these innovations in order to perform competitively and retain their funding. The end result was a number of virtually identical systems. Such evaluations are a powerful depressant on the risk taking that is essential in science. You can't get off a local maximum without going down in performance and thereby risking funding.

The MUC example illustrates some of the difficulties of trying to manage research by tightly managing its goals. Progress in science depends on many things. A very tight loop of incremental improvements can go only so far.

Consider this analogy: adding one rung to a ladder every year. That wouldn't be a good way to get to the moon.

Hazards of Peer Review   An alternative to evaluation metrics in applied research is to use peer review. From a perspective of fostering breakthroughs, John Seely Brown is critical both of peer review and of efforts to closely measure the results of research:

I think that we are trying too hard to measure the consequences of research, often counterproductively. DARPA has moved back from being able to fund research ecologies where there was trust in the program directors at DARPA and trust in the research community. Perhaps the ecology itself has not been that trustworthy. I don't believe that, but it is arguable that that is true. Universities have viewed and taken things as entitlements, which is a very serious issue.

NSF is driven by peer review to the extreme. No new ideas come out of peer review. That is a fundamentally conservative process. It is almost impossible to get really serious cross-disciplinary research started in that space. A lot of the ultimately successful frontier research sounds pretty crazy to begin with and would never make it through peer review.

When incremental methods stall, making progress requires trying something radical. The challenge in managing research funding is in assessing which areas and projects have the potential for breakthroughs. Since federal funding of research ultimately comes from taxes, funding organizations are always under pressure to be effective in their allocation of funds and sometimes to address emerging public concerns. In this way the common methods used by funding committees to govern research and responsibly ensure high quality projects tend to get in the way of breakthroughs.

Mixed Strategies in Federal Funding   Most of the research stories recounted earlier in this chapter used the approach of "radical research" or following a problem to its root. This style of research is rather free-wheeling—new disciplines may be added and goals may be revised along the way. By its nature, this style of research does not lend itself to careful planning. It changes direction according to the unanticipated obstacles that arise. Such research is difficult to evaluate closely along the way and difficult to justify to peer-review committees.

Radical research confounds conventional research management at the federal level. Its "as-needed" or "on-demand" blending of disciplines makes it an ugly duckling to single-discipline peer review in basic research. Its willingness to experiment to generate new knowledge makes it too unpredictable for metric-oriented applied research programs. Its freewheeling style—relying on the instincts and inspiration of very talented researchers—makes it a poor fit for management methods intended to work with all scientists, not just the brightest ones.

Robert Kahn worked as a program manager for DARPA for many years. Today he manages research programs and advises on science policy from his position as president of the Center for National Research Initiatives. He is familiar with the problems inherent in creating breakthrough research programs:

Agencies like NSF and DARPA often work together on programs, but tend to have different approaches. NSF is motivated by interesting research areas. They tend to work by consensus in the research community. That is one way to make decisions on what to invest in. You ask people what they want to work on and then fund them to do that.

When you are really working at the *frontiers* of knowledge it is hard to know what innovations will come except by finding bright people and funding them. DARPA does some of that. DARPA tends to look at big problems that would have broad impact if you cracked them.

In order for there to be an initiative there has to be something that makes sense to do, a need, and presumably an idea with the potential for fulfilling the need. It has to be a big idea. If the problem is just helpful to the community but it's not a big idea, it is harder to get funding from the research agencies because solving the problem then would just be seen as a matter of business.

I don't think that there is a cookie cutter formula for how to make initiatives happen on a national scale. Generally you need to get multiple parties involved.

It's usually desirable if one of the many research universities proposes it. Furthermore, the research agencies have to be comfortable with the plan. The NSF has done many initiatives usually around centers or cooperating centers.

Choosing which research to fund is a serious and difficult responsibility. Different agencies have developed different approaches, and no single approach works all of the time. Many of the research managers at NSF are career research managers. To protect against bureaucratic entrenchment and ensure high scientific quality of proposals, NSF and NIH (the National Institutes of Health) use a system of peer review in which proposals for research funding are evaluated by a committee drawn from the research community. This approach has the empowering effect of using scientific input from informed scientists. It potentially has hazards too, since people on the peer-review committee may themselves be competing for similar funds. Another hazard is that ideas can be driven more to the average because potentially brilliant proposals that are too radical for the taste of some members of a review committee are likely to be downrated. DARPA often gives program managers much more latitude to operate with less peer review, but gives them short-term appointments. Sometimes DARPA managers are highly motivated to leave a mark during their tenure and are willing to fund more radical proposals with more of a high-risk/high-payoff profile.

Reflections

Inventions can have three kinds of obstacles. The personal computer faced all three. It faced *creativity obstacles* for the graphical user interfaces, computer networking, and laser printing. It faced *innovation obstacles* in engineering and manufacturing. It also faced *change obstacles*, because it challenged how computers were used and how office work was done and because it had a huge effect on how people communicate.

Obstacles Long Forgotten
When radical inventions finally become widespread, they also become familiar. Clunky versions get replaced by better ones, and the faults and difficulties of the first ones tend to be forgotten. People who use later generations of an invention often don't know how things were before.

Electric Lighting   Today the light bulb is ubiquitous. When the light bulb was invented, however, it couldn't be sold in corner stores. People could not use light bulbs even if they could buy them. Houses didn't have the necessary sockets, switches, and wiring. There were no generators, no power companies, and no systems for distributing electricity. People who were interested in creating distribution systems disagreed about many standards. What should the standard voltage be? How should a plug look? Should the current be alternating, or direct? There were major obstacles to overcome before the light bulb could become a conventional household item.

Telephones   Imagine the questions that came up when the telephone was invented. How many phones are needed? There is no market for one telephone. Will they be used mainly for long-distance calling? If so, how many phones are needed per city? Is the telephone essentially a voice version of the telegraph? Is it much better? How many phones are needed in an office building? Is it worth installing phones just to talk to somebody on the next floor, or should phones be reserved for communications across several miles? Should there be many phone companies, or just one? If there are many phone companies, how do people subscribing to one phone company talk to people subscribing to another? How many operators will be needed if phones are given to everyone in a neighborhood?

After more than 100 years of evolution of telephone technology, these questions may seem quaint. Still, the evolution of the telephone is far from complete. Do we need many phones for locations—such as home phones—now that mobile phones are proliferating?[38] Movies from the early 1990s show people on the street talking into mobile phones that look like giant bricks. Each year since then, mobile phones have gotten smaller. Today it seems strange to see people walking down the street seemingly talking to themselves but actually speaking into phones. Will technology for subvocalizing[39] make this street muttering disappear? Will phones be surgically implanted?

Automobiles   When automobiles were introduced, they were for enthusiasts and wealthy people who could hire personal drivers and mechanics. Many details affecting usability needed attention, such as a safer and

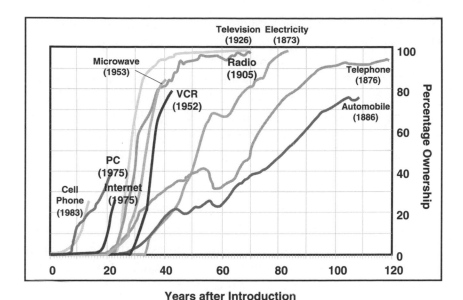

Figure 10.1
Percentages of Americans owning various technological devices. Used with permission of Gary Starkweather.

easier way to start an engine than with a manual crank. Tires were so unreliable that an enthusiast would expect to have at least one go flat on any weekend outing. There were no gasoline stations along the highways.

A big infrastructural change was the development of adequate roads. Not until the 1950s, many decades after the automobile was invented, was an interstate highway system built (to support trucking). Before then, a long interstate drive was much more of an adventure.

Railroads   In the beginning, each railroad line established its own track gauges. Cargo leaving one line to another had to be unloaded onto different trains because the tracks were different. This impediment to cargo transport was resolved when the railroads consolidated.

Adoption Curves and Tipping Points
Figure 10.1 shows technology-adoption curves for several radical technologies of the 19th and 20th centuries. Several things are notable. Some

technologies, such as the automobile, took decades to be widely adopted. Other technologies spread rapidly, going from under 10 percent to 80 percent market penetration in about 5 years.

What conditions enable some technologies to grow market share very rapidly and spread? One way to think about what enables rapid spreading is by analogy to wildfires. If someone tosses a lighted match on a pile of water-soaked leaves, the match will just fizzle out. In contrast, if someone tosses a lighted match on a pile of dry leaves under dry underbrush, the small flame can unleash a wildfire. The matches in the two examples are identical. The difference between the outcomes arises from differences in the context rather than in differences between the matches. Whether an invention takes off like a wildfire depends on both the invention and the context. The original Internet[40] spread fairly slowly when it was used mainly by academics for exchanging email and data files. One would not say that the "match" fizzled, but compared to today it did not go very far. The opening of the net to commerce changed the context radically, setting the stage for the invention of the World Wide Web.

Malcolm Gladwell uses the wildfire metaphor in his book *The Tipping Point*. His term "tipping point" refers to "that magic moment when an idea, trend, or social behavior crosses a threshold, tips, and spreads like wildfire." These phenomena of sudden and rapid growth also appear in the physical world. Epidemics take this form. Crystallization can occur quickly when a solution is super-saturated. This is called a *threshold effect*.[41]

Figure 10.1 suggests that technologies facing the greatest obstacles spread the most slowly. For example, the technologies that required the building of new infrastructures—tracks for trains, phone lines for telephones, gasoline stations for automobiles, and so on—rose more slowly along their adoption curves. Next-generation technologies can often spread more rapidly because they reuse the infrastructure pioneered by the first generation. For example, television reused production studios, transmitters, and advertising relationships that had been developed for radio. It climbed its technology-adoption curve more rapidly.

Strategies for causing tipping points are represented in both business and invention. In his books *Crossing the Chasm* and *Inside the Tornado*, Geoffrey Moore looks at business strategies for companies to get from the slow rising part of a technology-adoption curve to the faster-moving part. He shows the advantages of starting from niche markets. Success in niche markets establishes the usability of an invention and reduces the risks for bigger deployments. Opinion leaders point to the successes in niche markets and spread enthusiasm for a new technology. As the mainstream market becomes interested, the successful strategies focus on rapidly growing market share.

Understanding the dynamics of market adoption creates opportunities for invention. Opportunities are the flip sides of obstacles. If an invention is too expensive, the opportunity is to find a way to make it cheaper. If an invention is too large to carry, the opportunity is to find a way to miniaturize it. Opportunities are found by asking what qualities could make an invention more attractive to a larger and more demanding market. For many inventions, it is crucial to improve ease of use in order to reach a broad market.

Evolving toward Simplicity

Although the addition of new features in early versions of technologies tends to make them more complex, many technologies go through radical improvements toward ease of use before they spread to mass markets. The automobile is a good example. Climbing to higher levels of market penetration required climbing to new levels of reliability, convenience, and simplicity.

Although personal computers are still notoriously difficult to use and keep running, they have become simpler. In the same vein, today's upscale home entertainment systems, with video feeds from satellite systems, cable networks, digital video disks, video tape recorders, and video games, often require multiple button-laden remote controls. The controls for today's personal computers and home entertainment systems are about as primitive as the interfaces to early automobiles, with their dangerous cranks and unreliable tires. Their baroque complexities will be as bizarre to people in a few decades as the early automobile controls are to drivers today.

11

Cultures of Innovation

...the financial conditions necessary to think freely and deeply are hard to create.

—Jim Gibbons[1]

Client-Oriented and Patron-Oriented Cultures

Innovation cultures in a company are shaped by the interactions between its business perspective and its invention perspective. The business perspective focuses on revenues, markets, improvements to existing products, and known customers. The invention perspective focuses on new technologies and potential future customers.

The business perspective corresponds to the usual professional practices in engineering and design, where projects are funded through *clients* or customers. In a company, the business and marketing groups are the clients for the research and development groups. Clients exercise substantial control over choosing projects and want to grow their current markets. Project quality is determined by testing. The client model is oriented to the short term. It focuses its attention largely on the question "What is needed?" A client does not want to turn over control of a project to inventors, since that may lead to inventions that the client cannot commercialize.

The invention perspective corresponds to the traditional professional practices in the sciences and the arts, where project funding is provided by *patrons* or sponsors. When the patron model is used in a company, typically upper management rather business group management acts as the patron for research. Patrons exert minimal control over choosing projects. Funding continues if the patron is satisfied; project quality is

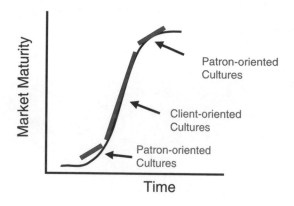

Figure 11.1
Dominance of patron-oriented and client-oriented cultures relative to a company's position on the technology-adoption curve.

judged by peer review. The patron model is usually oriented toward the long term and toward building a knowledge base for the exploration of new kinds of products and applications. The invention perspective focuses largely on the question "What is possible?" A patron does not want to exercise too much control, because it wants breakthroughs and options that go beyond the current market and challenge current thinking.

Figure 11.1 shows how the dominant orientation of the culture of innovation for a company tends to correlate with the position of its products on the technology-adoption curve.

When project funding for corporate research is mainly determined through a process of contracts with the business groups, the innovation culture is client-oriented. When project funding for corporate research is determined by upper management, the innovation culture is patron-oriented. When project funding comes from both sources, a mixed model is being used.

Stories

Before television, the main reward for excellence at sports was the implicit reward of a job well done. Television changed the way sports works. With television came advertising. Advertising is a way of monetizing stardom in sports by selling running shoes and other goods to those who follow the sports.

A career in science or technology is like a career in sports in that excellence matters and in that the activity by itself has limited intrinsic monetary value. You can't make a living just inventing. There has to be a way to take inventions to products. This is fundamentally why invention requires innovation. Taking inventions all the way to products is what creates value in the world and sustains the creation of further inventions.

A culture of innovation perpetuates its best practices through the stories it tells about itself. The stories tell "how our best people have succeeded." Legendary stories about the innovations of its heroes are passed on to new members of the community. These stories endure. In many cases the innovation stories shape innovation culture decades after they have happened.

In the film *24 Hour Party People*, the Tony Wilson character (a British television personality turned music mogul) quotes a line from the American film *The Man Who Shot Liberty Valance*: "When you have to choose between the truth and the legend, print the legend." Innovation stories in research laboratories are told and retold, eventually becoming legendary and heroic. The stories have power. Innovation stories convey attitudes and textures rather than a dispassionate and accurate history.

Client-Oriented Cultures

Fast-growing companies tend to be uninterested in finding or developing new or radical technologies. Confident in their current product and technology directions, they work to maximize their growth by focusing on their current business. Companies in this part of the S-curve use client models to manage innovation.

The client model is easier to succeed with in the near term. There is less immediate business risk, because the client already knows how to market the product and has established customers and branding. The needs for improving a product are known with greater clarity than the needs for a completely new market. Because of the short-term focus, invention under the client model yields incremental improvements to existing technologies.

Management Knows Best   As early as 1919, the Radio Corporation of America (RCA) had research facilities for studying long-range radio communication.[2] In the early 1930s, David Sarnoff reorganized RCA and gained control of the research staff of General Electric's vacuum-tube plant in Harrison, New Jersey and of the Victor Talking Machine Company in Camden. In the 1930s, the research program led to an early version of black-and-white television. Through the 1940s, RCA researchers worked in opto-electronics, high-frequency tubes, and acoustics.

An event in the history of a laboratory can be so dramatic that it becomes the new dominant myth. The best stories bubble to the top. Henry Baird encountered such a story when he worked at Sarnoff Laboratories[3]:

David Sarnoff made his name initially in radio and black-and-white television. However, the main stories I heard were from the early 1950s. There was a crisis because CBS had proposed an approach for color television to the Federal Communications Commission. It was based on a color wheel that spun at a high speed, superimposing color filters over black-and-white images. By synchronizing the illumination of the pixels with the spinning of the color wheel, a kind of color television was possible.

Sarnoff did not want a mechanical standard to prevail. He wanted color television to be all electronic with no moving parts. It should also be backwards compatible with the existing black-and-white television standard. This was an audacious goal. He decided that cost was no object. An approach was laid out and teams were assigned to the different parts of the task.

Researchers dropped whatever other projects they were doing and mobilized for the effort. They worked night and day to create a working prototype. The work continued through the early 1950s. Amazingly in 1953, the FCC reversed its approval of the CBS system. On the recommendation of the National Television Standards Committee, it approved the all-electronic version from RCA.

Sarnoff labs got a lot of patents for their approach. The licensing royalties from the patents on color television were so substantial that 20 years later, Sarnoff Labs was still receiving large funds from them.

The effect of this story on the culture of Sarnoff Labs was the idea that the role of management is to set difficult but achievable goals for research. The job of research is to accept those goals and to work creatively to meet them. So the lessons were that large crash projects work and that researchers work best in teams managed by senior technical people. Research was managed with milestones and deadlines. That culture persisted in the labs for at least 20 years, including the 10 years I worked at Sarnoff.

Over time this culture probably hurt the labs. Business needs were changing. Sarnoff's successors lacked vision, were risk averse, and found it difficult to give the labs a clear direction. Since then, the culture has changed at Sarnoff. Sarnoff is now part of SRI International.

Had Sarnoff not mobilized the laboratory, the CBS standard would probably have dominated color television for many years. The clarity of this goal gave great weight to the business perspective on innovation and was a defining moment for the company. The resulting culture of innovation lasted for many years, until later changes in the business position of the company were deeply understood.

A Moroccan Marketplace   SRI International is a scientific research institute in Menlo Park, California. Formerly known as Stanford Research Institute and affiliated with Stanford University, it was founded in 1946. SRI International, including its subsidiaries SRI Consulting and Sarnoff Corporation, employs about 1,400 people worldwide. Its website lists about twenty commercial clients and about twenty government agencies as clients, as well as several industry consortia and foundations. Jerry Hobbs is a computational linguist who was working at SRI International at the time of our interview[4]:

> The hero's journey myth at SRI is based on Doug Engelbart[5] and the invention of the mouse and modern computing in the 1960s. The myth is that a group holding a vision—working largely on their own—can change the world.
>
> Although Doug Engelbart's story inspired certain individuals, I don't think it had a huge impact shaping culture at SRI. We had a group here a couple of years ago that developed technology such as sensors to make the environment responsive and smarter. For example, they had a sensor in a refrigerator that would keep track of food going in and out. They worked with Engelbart, but for the most part, that myth no longer has much power at SRI.
>
> Research at SRI today mainly follows a different myth emphasizing the independence and responsibility of research groups. Each group is largely responsible for securing its own funding. SRI has been compared to a Moroccan marketplace where there's a bunch of independent merchants inside a single tent. The research is primarily driven from the bottom up. If not the bottom, at least one layer above the bottom. Each laboratory has somebody who maintains relationships with the funding people and makes sure that the researchers with new ideas get hooked up with the right funding people. Each group must attract its own sources for funding its work. Management can sometimes fund an effort like that or keep groups alive between grants. This interim support is very limited. Usually, when the funding goes away the people go away, or they go to work in other areas.

Curtis Carlson,[6] president of SRI International since 1998, says that successful projects require passionate champions and create "compelling value for real customers." Having champions to define and steer projects is crucial because management can never be smart enough to be the gatekeepers, deciding which projects to approve. With a broad range of clients, no one can know enough to pick projects across the board. Carlson asks "Would I be right 5 percent of the time?" According to Carlson, the job of management is not to set direction. It is to make sure that researchers are in the right jobs, and to support them with a team having complementary talents so that they can succeed.

The dominant culture at SRI relies on the initiative of individual researchers to acquire their own funding from outside the institution. In this framework, the role of management is to help make the connections. When an institution relies heavily on grants to fund research projects, it becomes client-oriented to the extent that its research managers must work to establish client relations. The direction of the program is strongly guided by the clients. In some cases where government funding has very-long-term objectives, the approach takes on characteristics of a patron model.

Demo or Die   The MIT Media Lab[7] was founded by Nicholas Negroponte and Jerome Wiesner in the mid 1980s. Seeing a convergence of computing, publishing and broadcast communications, Negroponte and Wiesner brought together researchers in a range of disciplines including cognition, electronic music, graphic design, video, and human-machine interfaces. At the beginning of 2002, the Media Lab had about 400 researchers (including graduate students). Owing to an economic downturn, the Media Lab had much more difficulty with funding later in 2002.

The Media Lab is a department of MIT's School of Architecture and Planning. The culture of the lab is one of bold thinking grounded in provocative project demonstrations. Funding of the Media Lab is a combination of grants and corporate sponsorship. A periodic Open House is a central part of the research culture that not only brings in sponsors, but also galvanizes and provides feedback about the lab's work. The Open House story is re-enacted periodically, and Pattie Maes sees it as central to the research culture[8]:

An important ingredient in the success of the Media Lab is really the fact that it's sponsored by industry. Every six months or even more frequently we have events for the sponsors where 200 to 1,000 people will show up and the sponsors have an opportunity to see all of the projects. We often have new technology for people to try when they visit for the day.

That's actually what we did with the Classified or the Markets system. We gave the 200 participants who came from industry three items at random from a collection of give-aways that we have—books by faculty at the lab, Media Lab umbrellas, watches, and mugs. We gave them three randomly chosen objects. We told them that while they were at the Media Lab for symposia and the Open House, they could enlist computer agents to buy and sell objects on their behalf. When the trading was over at the end of the day, they could walk away with the objects they were most interested in.

We set up kiosks in the atrium of the lab and gave them some fake money. The participants had a great time. The computer agents were bidding on their behalf while they were listening to the sessions and talks in the auditorium. We gave participants pagers so that if an agent made a deal on their behalf, it could page them. It would tell them on the pager which of their items they had sold or what they had bought and where they had to meet the other party to do the trade.

Creating real systems is a tremendous amount of work, building them and seeing that they function properly. In this case my students were hacking the system all night long to get it working. The system was ready to use only three minutes before the sponsors walked out into the atrium. I was literally giving the talk about how to use the system in the front of the room. At end of my talk, or almost near the end, I got the okay from people at the back saying "It's working!" Otherwise I would have had to tell them to wait another hour to use it.

In the tenure race of academic appointments, the mantra over the years has been "Publish or perish." At the Media Lab, thinking and publishing is not enough. Its culture is derived from design and architecture where building prototypes and giving demos are essential ingredients in learning about what works. The Media Lab mantra is "Demo or die." When you build something, you see the problem more completely and you can see aspects of how it works or how it could be used. For this kind of insight, the unaided imagination is often not powerful enough. The innovation stories at the lab surround the giving of demos at events like the Open House.

A second reason for demos is that they are useful for conveying new concepts to sponsors. In communications based only on posters, slides, and papers, the listener or reader must duplicate the imagination of the inventor. With a demo, there is more to touch and see, and the potential becomes much easier to understand. Demos are concrete, and communication becomes easier.

At the Media Lab, the demo culture is used as part of effective communication with clients. The Open House is a vehicle for showing a lot of new stuff all at once. This is the niche of the Media Lab, where the

"Demo or die" myth is re-enacted on every project and is acted out in parallel across many projects with every Open House.

Shifting from Client Models to Patron Models When the niche of a company changes, it needs to reexamine its culture of innovation. Gordon Moore, co-founder and former chairman of Intel, guided the company as a dynamic high-technology organization with no central research organization. In 1996, in a personal retrospective[9] on research, he argued that this was an important principle:

Intel operates on the Noyce[10] principle of minimum information. One guesses what the answer to a problem is and goes as far as one can in a heuristic way. If this does not solve the problem, one goes back and learns enough to try something else. Thus, rather than mount research effort aimed at truly understanding problems and producing publishable technological solutions, Intel tries to get by with as little information as possible. . . . Another advantage to operating on the principle of minimum information: the company generates few spinoffs. Because it does not generate a lot more ideas than it can use, Intel's R&D capture ratio is much higher than Fairchild's ever was. (p. 168)

Moore's position is that of an engineer, firmly in Edison's quadrant. He had seen other companies in the semiconductor business invest in research and fail to capture as people left and formed unintended spinoffs. These observations led him to focus strongly on controlling research to prevent wasted effort.

Moore's attitude toward research was dominant during the rapid growth of Intel's business of making chips for personal computers in the 1990s. On this part of the technology-adoption curve, a company does best by focusing on growing its market and not getting distracted by other investment opportunities, including those that could be created by research.

No niche lasts forever. By 2002, Intel faced saturation in the market for microprocessors for personal computers, and its attitude about research began to change. It expanded its internal research laboratories and developed a program of outsourcing research to "lablets" at other institutions.

In the story from the 1950s about Sarnoff Laboratories, the client-oriented culture enabled the laboratories to play a major role in establishing the color television standard and readying RCA for a major expansion in its business. Decades later, when television evolved to become more of a commodity, the business niche of RCA changed and management was less able to provide a clear direction for the laboratories. The client-oriented research culture didn't fit as well.

While this book was being written, research at General Electric's Global Research Center was making a similar shift from a client model back toward its roots with a patron model.[11] The labs were initially set up by Charles Steinmetz in 1901 to do fundamental research. In the 1990s, GE CEO Jack Welch insisted that the researchers petition the business units for funds. He believed that the research organization acted "too much like Bell Labs" and that the business units would not pay for projects that merely satisfied scientific curiosity. Since Jeffrey Immelt became CEO, GE's industrial businesses have struggled in the slow economy and profits for its insurance businesses have suffered with the terrorist attacks. United Technologies and Royal Philips Electronics are also stepping up long-term research. The *New York Times* quoted Immelt as saying "Superior technology is the only route to consistent growth without declining margins." This new attitude has led to significant shifts for funding research. Funding for long-term research grew from $28 million (10 percent of a $286 million budget) in 2000 to $104 million (30 percent of a $349 million budget) in 2003.

In the Intel, Sarnoff, and GE examples, the innovation culture shifted from a client model toward a patron model when the markets saturated and the company believed that it needed more breakthroughs. Companies shift to patron models when they believe that the current businesses need to evolve to sustain growth.

Businesses and societies that don't anticipate or create change eventually pay a heavy price. This issue is at the core of Clayton Christensen's book *The Innovator's Dilemma*. Christensen's case studies are about companies so pre-occupied with short-term thinking and extracting value from their existing high-volume products that they don't anticipate what is coming. Files of business case histories are filled with stories of companies tending to the short term and being swept aside by somebody else's innovation.

Patron-Oriented Cultures

If a client greatly narrows where to look (for example, by limiting the view to incremental improvements of current products), breakthroughs are left out of view. A future business opportunity will go unnoticed or be discarded because it looks so different from current products. In a patron model, invention is driven more by the dictates of science than by the dictates of current markets. Because of its long-term focus, invention under the patron model is less constrained by a company's current products, and is more likely to lead to breakthroughs.

Benign Neglect Bell Laboratories, founded in 1925 and renowned for the invention of the transistor and for six Nobel Prizes, was a power-house of inventions created under sustained and stable funding for the AT&T monopoly. At its peak, it employed 30,000 scientists and engineers in 29 countries.

Bell Labs no longer exists today in the same form or on the same scale. In 1984, when AT&T was broken up into the regional companies that came to be called "Baby Bells," the labs were divided and Bellcore was created as the applied research arm of the regional companies. In 1996 AT&T decided to spin off its system and technologies business into Lucent, which got the lion's share of Bell Labs. Some researchers moved to Bellcore. In 1997, Bellcore became Telecordia Technologies and was acquired by SAIC, a large systems-integration company. As Lucent continued to spin off large pieces, Bell Labs broke up further.[12] When Henry Baird worked at Bell Labs, he encountered two different innovation myths, one told by the physicists and one told by the computer scientists. Baird:

The physicist's myth was based on the story of the invention of the transistor by William Shockley and several others. Everyone at the time knew that a replacement was needed for vacuum-tube amplifiers. It should be cooler, smaller, and more reliable. No one knew how to make one, but semiconductors were in the air. According to the myth, a group of three geniuses, working more or less undisturbed, pursued the quest with little prodding or interference from management. Out of their scientific interactions, the transistor was born.

The basic lesson was that the research approach should be figured out by a team of two to seven people. Management should set the general direction, but then get out of the way. Researchers know best about how to pursue the research.

The cultural effect of this myth on Bell Labs led to a profoundly different attitude toward research than at Sarnoff Labs. Management has a role in suggesting goals, but then its part in guiding research is finished. It was expected to get out of the way, pay the bills, and let the researchers get the job done. Project reviews, milestones, and so on were not encouraged. Compared to the model at Sarnoff, this led to much more freedom and initiative for the researchers.

During the long period when Bell Labs dominated science, AT&T was essentially a monopoly in the communications business and was looking for new businesses to enter. Because of its monopolistic position, it resided in the mature market phase for a long period of time and was able to sustain a patron-based culture of innovation.

In an interview conducted by Ernest Miller, Joshua Lederberg described a similar attitude about managing academic research at Rockefeller University:

I think it's the job of management to define research goals but to be very cautious about centralizing control of the means, how these goals will be pursued. If you can convey your goals to the people who are actually confronting nature in the laboratory and get them to internalize them, that's about as far as you ought to go. They'll be far more capable of effective research when they are using their own imagination and direction and relying on the observations they make from day to day to guide their efforts to meet those goals than they would be in following the instructions of any central research manager.

Researchers Know Best   Bell Labs has been large enough and long-lived enough to have many important successes and innovation myths. A second important myth from the lab concerns the launching of the UNIX operating system. UNIX is an operating system that was developed in the 1970s under the guidance of Dennis Ritchie and Ken Thompson. Reliable, powerful, and simple, it swept through the computer science community like a storm. Today, versions of UNIX are used in many of the computers built by Sun Microsystems, Hewlett-Packard, IBM, Apple, and other companies. The LINUX operating system for PCs is a more recent addition to the UNIX family tree.

In the 1970s, AT&T had hoped to get into the computer business. Before the development of UNIX, there was a large, multi-year effort at Bell Labs to create an all-purpose operating system called MULTICS. Although MULTICS was the seedbed of some interesting computer science, the project was hopelessly complex. When the huge MULTICS consortium failed, it was decided that never again should Bell Labs attempt to create software as complex as an operating system. This failure set the context for the UNIX story. Henry Baird:

The computer scientist's myth is about UNIX. It starts with the failure of MULTICS, which was an embarrassment to a whole generation of managers. In this climate, two researchers who needed a reliable and simple operating system for their research began working on one essentially in secret. They found an old

PDP-11 and I heard that they hid out in an attic room to build their operating system. They were geniuses and there were occasional fights and difficulties—but through it all, their manager kept their project secret until it was done. According to the myth, it never showed up in the annual reports; it was never given milestones; and it was never accounted for. When it was done, UNIX was released to the world and swept through the community as a roaring success.

The lesson of this story was that undirected research works. In the culture, it says that individual researchers both know best what to look for and what is needed. Given the MULTICS background, the example said that managers were incapable of deciding what is possible or what is worth doing. The way to manage research is thus through benign neglect. Projects are voluntary. Researchers are hired for their personal initiative and the power of their central nervous systems. The job of management is to pay the rent and stay out of sight.

The leaning toward the innovation perspective in the UNIX story did for computer science at Bell Labs roughly the same thing that the transistor story did for the physicists. The success of UNIX came at a time when technologies for computing and telecommunications were merging and gave AT&T increased visibility and credibility in the computer community.

Shifting from Patron Models to Client Models   In 1949, the aluminum industry was shifting out of wartime production. It was challenged to recast its "battle-tested metal" for civilian applications and to produce on a much larger scale. Alcoa formed a Technical Committee to set priorities within the research program. This committee was intended to minimize duplication of effort and to ensure that research would be directed to the most pressing problems. This prioritization led directly to a shift from a patron model to a client model. Researchers were moved from project to project following shifting market priorities. This caused a significant loss of independence for the laboratories. In *R&D for Industry*, Graham and Pruitt noted how sustaining this policy affected the fortunes of the company over time:

To some it might seem a trivial matter, but by removing the crucial allowance for uncertainty in longer-term and exploratory work, such a system could, over time, drive a research program into shorter-term, more predictable work. Such R&D would meet its targets and reliably deliver what could be foreseen, but that style of resource allocation would foreclose the unforeseeable, the serendipitous, and probably the most fundamental innovations. (p. 306)

The shift from a patron model to a client model has happened for many businesses. At IBM, the peak time for the patron model was the period from the 1960s through the early 1980s, when the "computer giant" dominated the computer landscape. IBM is a large company with many businesses, and its diverse products occupy many different points on the technology-adoption curve. With the advent of the personal computer, IBM saw Microsoft and Intel emerge as driving forces. IBM became just one of many suppliers. IBM also faced increased competition from Sun and Hewlett-Packard and from other companies producing "server-based" products. Consequently, IBM realized that it no longer dominated all areas of computing. IBM developed a sense of urgency to *focus* its research resources and a greater belief that management knew what was needed. This led to a shift in the balance toward a client model.

In the beginning of 2003, IBM announced its intention to shift its research culture even further toward a client model. The approach involves allocating a portion of the research organization to focus specifically on customers' problems in conjunction with IBM's growing consulting business. This move is intended not only to help IBM understand its customers but also to help the research organization, which has traditionally focused on IBM's technology business, to find ways to support its service business. Since IBM's consulting business was entering a period of rapid growth, this shift toward a client model again follows the predictions of figure 11.1.

In a similar way, Lucent has guided Bell Labs toward the short term and toward applications in Lucent's communications business. Lucent's difficulties were part of the economic collapse of the communications industry.[13] The labs have become more focused. Turning toward a client model, research projects now have "relationship managers" and champions to coordinate projects with Lucent's business units and also increased contact with its customers.

Cultures That Mix Client and Patron Models

From the perspective of evolving businesses, neither a pure client model nor a pure patron model is reasonable. Reflecting on the need for periodic renewal, Graham and Pruitt consider the experience at Alcoa:

The appropriate task of managing R&D is not to achieve either complete harmony between the laboratories and the corporation as a whole or unity with the scientific community at large; rather, it is to maintain constant tension between the two. A creative tension must also be preserved with the R&D program, between the operating need and the new opportunity, between the immediate and the long-term, between the fundamental and the applied. Managing R&D is finding the elusive balance point on these several different axes. (p. 503)

Many research cultures mix the client and patron models with a portfolio of projects. It is common for research laboratories to seek a balanced profile to accommodate both current and future prospects for a company. For example, at a forum of computer professionals, David Waltz of the NEC Research Institute characterized that institute's work as 30 percent long-term research, 25 percent "targeted science" (3–5 years out), and 45 percent advanced development, versus 100 percent "basic research" when the institute was founded. Dick Waters of MERL (Mitsubishi Electric Research Laboratories) and Dan Russell at IBM described a rule of thirds—one-third short-term, one-third medium-term, and one-third long-term. When a portfolio approach is executed well, it mixes a continuous and fairly predictable stream of short-term innovations with episodic and less predictable long-term breakthroughs. This can create value for both present and future customers.

Making a Positive Impact   Dan Russell heads the User Experience Lab at IBM's Almaden Research Center near San Jose. He has also headed a group with approximately the same name and mission at Apple's Advanced Technology Group (ATG) and at Xerox PARC. He is in a good position to compare these three corporate research cultures.

IBM is known for making and selling computer systems. Its primary businesses include computer hardware, software, and information services. It employs about 300,000 people worldwide and had $85 billion in revenues in 2002. IBM has eight research labs around the world, with about 2,000 researchers.

Apple is also known for the computer systems it makes, primarily personal computers. When Russell worked at Apple in the early 1990s, annual revenues were about $4 billion and there were about 10,000 employees. ATG, which no longer exists, had about 100 employees.

Russell worked for Xerox PARC more than once, most recently in the late 1990s. At that time Xerox had about $18 billion in revenues, mostly in manufacturing, selling, and leasing copiers and printers. It had about

100,000 employees and five research centers around the world. About 250 researchers worked at Xerox PARC. Russell[14]:

Every company has a set of memes—the deep ideas embedded in the culture that everyone takes for granted. IBM is really big. One of the memes is that because IBM is so big, it expects *big* wins. If you come up with an idea that's going to sell $10 million next year, there is little interest. If it's $100 million, management will start to listen to you.

In the course of its years of operation, IBM has been in almost every possible business. At the IBM museum back in Armonk, New York, there is an IBM meat slicer, so it's been in the food products business. A consequence of having tried so many businesses is that it's very difficult to reenter a market where we have failed once.

In this research culture, only *impact* matters. You are told this time and time again. Of course researchers publish papers, but what *really* matters is that you impact something that sells. "Your research is great. Thank you, doctor. I love the analysis." But if you have an influence on a big product that's made some big bucks, *that's* good.

Most of the research funding at IBM comes from a "tax" on the divisions and is pre-allocated for the research budgets. However, each laboratory *also* has to go to the product groups and obtain about a third of its operating budget from negotiated contracts. As a lab manager, I have to go to the Lotus or DB2[15] product divisions to negotiate a joint program funding agreement. We are accountable for doing work for the product division and they pay us a fee. Part of my job is figuring out how to get a third of my lab's operating budget in this way. A consequence is that the laboratories are very formal about connecting with the IBM businesses.

The way you get status inside the IBM research culture is to make a big play. For example, somebody might invent a new way of indexing databases so that re-indexing time is cut in half. Doubling the speed of a product makes IBM more successful. Making such an impact for the corporation makes you win as an innovator.

Over time the culture of IBM's research division has changed. My colleagues who have been at IBM longer tell me that there used to be more of a luster of science role. Big scientific ideas would give you status. A Fields or Nobel medal was a great thing. Publishing in *Science* or *Nature* or creating a new branch of mathematics such as fractal theory would give you status. Now, that's not as important. It's still nice to redefine scientific areas, but in term of advancing your internal career, and in terms of the people that we hire, it's more important to have a business impact.

Apple is an interestingly different company from IBM. Apple sells *integrated* hardware/software systems. They own the means of production. They control the entire design from soup to nuts.

This means that they can create an integrated user experience like nobody else can. IBM can't do that. For example, it does not control the booting software or all of the intermediate software. Other companies do. The main meme for Apple is "We are insanely great. We do elegant, beautiful work."

The research culture at Apple is about inventing new things that *nobody else* will think of. As they say, it's like shooting for a duck. You don't aim where the duck is now, but where it will be. At Apple you look at the technology trends, see where things will be, make a big bet and invest everything there.

For example, when QuickTime[16] came out everyone thought "What a waste of time." It had postage-stamp-sized windows and a slow frame rate. But Apple knew Moore's Law.[17] They knew that processor speed was going to increase. They invested early and shot exactly where they should.

Organizationally, ATG was one lab at the corporate center. This meant that you could go to the gym and talk to the people who designed the basic functions in the Macintosh operating system and get them to make small changes. Having everybody close together made the connections within the company very simple.

You got status at Apple by shipping a product. "That's a great demo. *Ship it!*" "Ship it!" was a phrase you heard a lot at Apple. There was a lot of interaction between research and the production group. We had an active shuttle program where we would send people back and forth between research and production. It was amazingly effective.

The social reality there was that the products have to be cooler than the competition's products. There was deep knowledge that elegance matters in design. Everybody at Apple knows that without question. Your product has to be interesting, novel, and inventive. The average age at Apple was pretty young, probably in the mid to late twenties. A side effect is that there are all these 20-year-old programmers who will basically kill themselves getting something to work.

I was at PARC before it became a separate company from Xerox. PARC stood in an interesting relationship to the rest of the corporation. By contrast with computer companies where you just walk down the hall to talk with the developers of the operating system, there is no obvious way to do that at PARC. That's both good and bad.

Not having a close coupling with business gives research *enormous freedom* to create new ideas. The problem then is taking the ideas back to Xerox. Technology transfer wasn't PARC's problem—it created companies like Adobe, 3COM, etc. They made money, but not for Xerox. PARC sent a signal but Xerox couldn't hear it. Fundamentally, there just wasn't a receiver.

The research culture at PARC has a high ideal. PARC researchers want to be the best of breed and a national resource. PARC is where the great ideas come from. PARC has more of an attitude of "We'll build the greatest stuff. People will come and recognize it." And that's effectively what has happened.

The bottom line in all three cultures is that impact matters. The ease in making an impact on the corporation was very different in the three cases. At Apple it was easy to connect with developers and influence products because you could just walk down the hall. At IBM it's a lot harder to maintain relationships with product divisions because the company is much bigger and more spread out. At PARC it was particularly hard to connect with the product groups because PARC was set up to do breakthrough research and that doesn't have an easy impedance match with short-term-oriented product groups.

The skills and training for working under a patron model differ from those for working under a client model. Researchers focused on answering "What is possible in the long term?" must spend substantial time beyond their specific projects building up intellectual capital for their work. This means spending time reading, thinking, and tinkering rather than talking to customers. Inventors focused on answering "What is needed in the short term?" need to spend substantial time coordinating with business units, exploring the market, and finding and cultivating the next client.

Keeping a portfolio of projects gives an organization some capacity not only for addressing both short-term and long-term needs but also for shifting the balance of effort between a short-term focus and a long-term focus. Such organizations recognize that different skills and aptitudes are needed for addressing short-term and long-term considerations. There is room for excellence and creativity in the long term and in the short term. Inventors who excel at short-term problems, with an engineering attitude, must be adaptable and must be good listeners. Researchers who excel at long-term problems tend to orient themselves to fundamental questions. They find that too much interaction with "what the customer wants today" is a distraction from tackling the deeper issues. By mixing the patron and client models, a business can institutionalize some of the creative tension between the business perspective and the invention perspective.

Reflections

Why Sustained Innovation Is Difficult to Manage

One metaphorical way to understand the fundamental tension between the business perspective and the invention perspective is in terms of clocks and compasses. Figure 11.2 shows two clocks and two compasses: a clock and a compass for invention, and a clock and a compass for business. The invention compass points to the next invention; the corresponding clock indicates when the invention will be ready. The business compass points to the next market; the corresponding clock indicates when the market will be ready. The invention clock is geared to progress in science and engineering. Inventions are created through the efforts of science and engineering and not by the wishes of science fiction fans. An

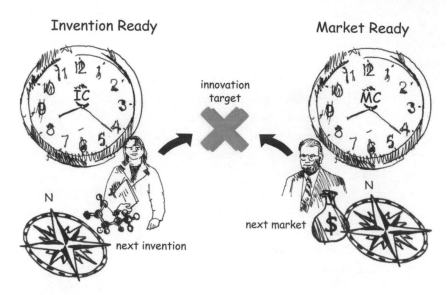

Figure 11.2
The invention clock and compass indicate what *inventions* will be ready and when. The business clock and compass indicate what *markets* will be ready and when.

invention clock can sometimes be speeded up by adding more resources. Its speed depends on the accumulated knowledge, instrumentation, and the maturity of scientific thought in a particular area.

The business clock is geared to the readiness of markets. Markets are not shaped by just one or two consumers or by the imagination of a marketing department. Markets need to be developed and follow technology-adoption curves. Although adding resources can sometimes help markets to form, business people do not completely control the market clock.

Similarly, the invention compass and the business compass point in the direction of the next opportunities. Innovation happens when *both* compasses point to the same opportunity and both clocks say that the time is right. This point is when the right invention reaches the right market at the right time.

The differences between these two perspectives reveal the underlying reason why sustained innovation is difficult to manage. In essence, the management of innovation must cope with two sources of uncertainty:

technology uncertainty and market uncertainty. Inventors and scientists are fundamentally engaged in managing technology uncertainties. They focus on technological feasibility, performance of technology, technology trends, and breakthroughs. Business people are fundamentally engaged in managing market uncertainties. They focus on value, markets, profits, customers, and market trends. Both perspectives are needed in managing innovation.

An aid to bringing these perspectives together is to use a set of intermediate concepts in a business model.[18] A technology business model is jointly constructed by inventors and business people and integrates the two perspectives. Creating a business model entails identifying both a possible market and a proposed offering. It requires articulating the attributes of the offering. It can entail identifying a value chain—a series of contributors to delivering the offering. It also entails finding a way to get paid. And it entails finding a way to sustain the business model.

Because a business model integrates technological and business perspectives, it requires the participation of inventors and business people. Not all inventors can work with the concepts of a business model, nor can all business people. The good news, however, is that these concepts tend to be accessible from both perspectives and provide a starting point for exploring new possibilities. The collaborative combination of perspectives—business and invention—is exactly the same "dance of the two questions" that we encountered earlier in the context of Pasteur's quadrant. It is in this quadrant that knowledge about what is needed inspires foundational science and invention, and knowledge about what is possible inspires a search for applications and markets. At its core, Pasteur's quadrant is about the dual-space search.

Experiment, Adapt, Iterate

In *Open Innovation*, Henry Chesbrough refers to two kinds of errors in developing new technology businesses. In a type 1 error, the business strategy assumes incorrectly that there is a market for a new technology when such a market does not actually exist. In a type 2 error, the business strategy assumes that there is not a market for a technology when such a market does exist.

Organizations that have learned to embrace the different perspectives learn to synthesize new directions from these conflicts. Missed opportunities or type 2 errors occur if the business people know of a market need but don't imagine that a solution is possible. ("We didn't know you could do that!") Another kind of missed opportunity occurs when the inventors know of an invention but don't realize that there is a market for it. ("Who would have thought that anybody needed that?") When the perspectives are brought together constructively, unsuspected opportunities for innovation can be co-created. The trick is in knowing how to search for the right technology and the right market. For new inventions, this search is best carried out as an iterative exercise.

One of the first questions asked from a business perspective is "What is the size of the market?" Failing to consider this part of the search is one place where a type 1 error is likely. The next questions are "How much of the market is accessible?" and "What will be our portion of the profits on this?" The *accessible* market is that portion of a market that *this company* could actually have, with its marketing reach, its manufacturing capacity, its competition, and the regulations that characterize its niche.

When an invention is just one part of a larger system, a company can expect to participate in the profits according to its contribution to the overall value. For example, suppose there is an existing $100 million market. Suppose that 30 percent of that market is accessible to the company, that the innovation contributes 10 percent to the overall value, and that profits after expenses would be 20 percent. The estimated value of the research would be no more than $600,000, which may not be enough to cover costs of research and development.

When the evaluation logic is limited to this form, it can seem that almost no invention is worth pursuing. This line of thinking overlooks the fact that new technologies can create new markets.[19] When that occurs, different methods for estimating value are needed, and different questions. This is exactly the point at which it is easy to make a type 2 error. Who are the potential customers for the new invention? How many of them could we serve? Why do they need the new invention? How is this need satisfied now? What would they be willing to pay for it? This alternate line of thinking can lead to estimates of market size more appropriate for new technologies.

Except for the questions bearing on market size, these questions are essentially what is known as "Heilmeier's Catechism,"[20] a method proposed by George Heilmeier that is widely used for prioritizing R&D projects. The catechism guides decision making by asking six Descartes-style questions:

What are you trying to do? (No jargon, please.)
How is it done today and what are the limitations of current practice?
What is new in your approach and why do you think it can succeed?
If you are successful, what difference does it make? How do our customers get paid?
What are the risks?
How much will it cost? How long will it take? What are the mid-term and final exams?

Several variations of Heilmeier's Catechism are in circulation. Some start out by asking "What is the problem and why is it hard?" One asks "How do we get paid?" and "What is the risk reduction plan?" Instead of the "exam" questions, some variants ask "How will intermediate results be generated and how will you measure progress?" Heilmeier's Catechism has been a part of the culture of evaluating project proposals at DARPA and elsewhere for many years.

A process of iteration treats the creation of a business model as a dual-space search, looking for a combination of a market and an invention. In effect, it employs the clocks and compasses from both the business viewpoint and the invention viewpoint. Small-scale experiments—essentially "marketing and technology probes"—can be conducted. Investments are made a little at a time, and information is gained at each stage. The information being sought is about what technology works and what nascent markets may exist.

This incremental approach is not really new. It is exactly the approach used for years by venture capitalists. Each round of funding is another iteration. What is new is applying these ideas to manage innovation in the context of research organizations and existing companies. It is well known that the budget is much larger at each successive stage from research prototype to product. Research organizations themselves have neither the budget nor the experience to carry out extensive market probes. In *Innovator's Dilemma*, Christensen recommends creating a separate and independent business for any radical technology. The most successful spinoff organizations from research also go through a transition to new management and venture-capitalist funding to use this model.

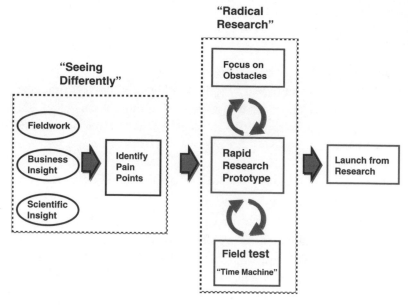

Figure 11.3
A process for reducing risk in radical research.

Heilmeier's Catechism is a prescription for evaluating new projects on the basis of their capacity to reduce risk. Figure 11.3 shows the main elements of an approach to breakthrough research in which the risks and the benefits are both very high. The first block in this figure recasts the theme of "seeing differently" in terms of the insights that inspire a project in the first place. Insights can arise either from business ("What is needed?") or from science ("What is possible?"). Augmented by field-work, radical research must identify an important problem (here called a pain point). Next comes the activity of radical research, the multi-disciplinary pursuit of a solution (or a "leverage point"). When obstacles arise, new disciplines may be brought into the mix. When a candidate solution is at hand, the next part is to validate the proposition. Does the new technology actually alleviate the pain point? Does it effectively solve a future problem for future customers? To answer the latter question, it can be effective to use a "time machine." A microcosm of the future is constructed to field test real people using the prototype solution. Sometimes groups in advanced development take part in carrying out and evaluating the field test. New insights may arise from watching people learn about and try to use the new technology. When the field test is promising, business questions receive renewed attention. What is the economic value proposition? What are the manufacturing costs? What is the expected market growth? By this point, the center of activity for the project has moved beyond research.

Timing matters. New technologies become new businesses when all these factors are in alignment. As Chuck Thacker said, "You can't build railroads before it is railroad time."[21] When "railroad time" comes for a new technology, the best-prepared companies are in the best position to take an early lead.

Big successes in innovation are rare. Success requires having enough passion to surmount all the obstacles that arise. For new inventions and markets, patience is needed to go through several iterations in developing a workable business model. When a great innovation succeeds it is worth celebrating. We ask "How did they *do* it?" The innovation stories from research institutions tell us how something both novel and needed came into the world.

12

Innovation at the Crossroads

Because of its nature, the industrial laboratory is an institution suspended between two worlds—that of industry and the marketplace, on the one hand, and that of the scientific professions, on the other.
—Margaret Graham and Bettye Pruitt, *R&D for Industry*

The Dynamics of Combining Ideas

The ideas behind tomorrow's breakthroughs are scattered throughout the innovation ecology. They are in universities, government granting agencies, venture-capital firms, corporate research laboratories, and other institutions. The ideas ride around in people's minds and also in magazines, newspapers, professional journals, and books. New ideas are generated as people are inspired. Ideas bump into each other when people discuss them over the lunch table. But the right ideas must be brought together and combined before breakthroughs can happen.

Ideas are seldom combined easily to create breakthroughs. Some settings offer richer possibilities for creating and combining ideas than others. When people from the same field meet, the opportunities are limited for radical combination because they all have the same background. When people from distant fields meet, ideas don't cross-pollinate without substantial effort. Efforts must be made to bridge the different technical vocabularies and build common ground before people can understand each other deeply. This is not true only of academic settings. It also happens when people in research, product development, and marketing come together.

Radical research encourages effective combination of ideas. People with different backgrounds come together and work on an important

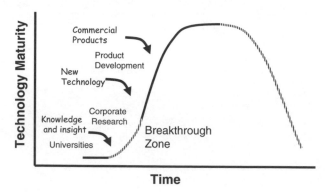

Figure 12.1

problem with patience, motivation, and common purpose. This creates both the opportunity and the context for breakthroughs.

Jumping the Two Hurdles

The breakthrough zone is the part of the innovation ecology from which most breakthrough technologies have historically arisen. Figure 12.1 (which repeats figure 1.2) is approximate, but it shows some major patterns.

Universities and federal funding agencies have largely been responsible for basic research. Basic research aims at increasing fundamental knowledge, rather than addressing specific problems. At the other end, product-development teams create products. They do not create basic knowledge and do not begin their work until technologies have been invented and proven. Work in the middle zone—done largely in corporate research laboratories—has led to most of the breakthrough technologies. Why is this the case?

One way to understand why breakthroughs cluster in the middle is to consider the two fundamental questions that are at the core of innovation: "What is possible?" and "What is needed?" Before something becomes an innovation, both questions must be answered. Each of these two questions is a hurdle to innovation. Suppose that somewhere in the innovation ecology a group is working to create a new technology. If the group is working on basic research, the question "What is possible?" is at the core of its work. The group's research life is organized to do science and to create new knowledge. Developing products is a long way off in

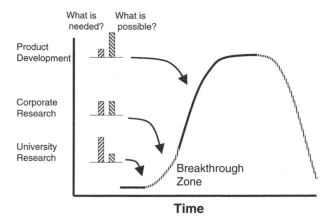

Figure 12.2
The bars at left represent the amounts of energy required to address the questions "What is possible?" and "What is needed?"

the distance. The group's research activities are neither driven by nor informed by a sharpened sense of what is needed.

Figure 12.2 illustrates the "energy level" needed in each region of the innovation ecology to address the two questions. In the region of basic research, the "What is possible?" bar is short, meaning that relatively little energy is needed to address that question The activities and colleagues in this region address "What is possible?" all of the time. At the same time, the "What is needed?" bar is very high for this region. Much more energy is needed to address that question, because basic research scientists are not experts on current or future needs. Addressing the "needs" question is difficult for them.

Near the top of the graph—in the product-development region—the opposite situation holds true. In this case, the group is much closer to markets. Its work on incremental product improvements is typically driven by the question "What is needed?" Addressing the "What is possible?" question is very difficult.

In the middle part of the curve, which represents corporate research labs, both questions appear again. Here the question "What is possible?" is of medium difficulty, since some members of the group or their colleagues are engaged in basic research. Similarly, the question "What is needed?" is of medium difficulty, since at least some members of the

Figure 12.3
In a university or a basic research setting, the hurdle corresponding to "What is possible?" is very low and the hurdle corresponding to "What is needed?" is very high. Very few jumpers make it over both hurdles.

group are connected to marketing or product development. Crucially, neither bar is of commanding and inhibiting height.

Suppose that the people in an innovation group are like a track team that must jump over two successive hurdles.[1] The greater the height of a hurdle, the smaller the number of jumpers that can jump over it. For the group performing basic research (figure 12.3), many of the jumpers can easily clear the short "What is possible?" hurdle. Very few can clear the tall "What is needed?" hurdle. Consequently, when all the jumpers have tried, very few will clear both hurdles.[2] By analogy, very few break-through technologies are created in this region. Similarly, in product-development (figure 12.4), many jumpers can clear the short "What is needed?" hurdle but few can clear the tall "What is possible" hurdle. Again by analogy, very few breakthroughs are created. In the corporate research setting (figure 12.5), both hurdles are of medium height. Getting over the medium first hurdle is a bit harder than clearing a short hurdle, but still many jumpers make it. Getting over the medium second hurdle is also easier. Consequently, *many more jumpers clear both hurdles*. By analogy, many more breakthrough technologies come out of this region.

Figure 12.6 summarizes the overall thought experiment. The analogy predicts that the middle region or breakthrough zone will be the most productive because neither hurdle is as imposingly high as in the other two regions.

Figure 12.4
In a product-development setting, the hurdle corresponding to "What is needed?" is very low and the hurdle corresponding to "What is possible?" is very high. Very few jumpers make it over both hurdles.

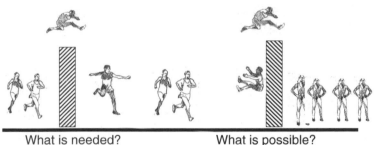

Figure 12.5
In the corporate research setting, both hurdles are of medium height and many more jumpers make it over both.

The challenge facing the innovation ecology is about finding ways to sustain the conditions that create breakthroughs.

Stories

Open Innovation
Open innovation (a term coined by Henry Chesbrough[3]) consists of strategies by which companies can acquire technologies they need and exploit technologies they have developed.

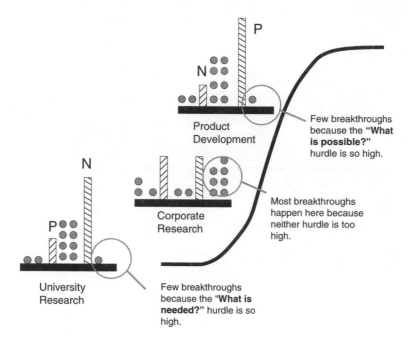

Figure 12.6
The breakthrough zone produces the most breakthrough technologies because more groups can get over both hurdles.

Closed versus Open Innovation   Ideas and inventions can come from many places. No company can hire all the smart scientists and engineers that produce technologies relevant to its business. The world's smart people are scattered about. They are not all members of a single team or corporation. They can be found in other companies, institutions, and countries. Today there are few knowledge monopolies.

The concepts of open and closed innovation are not entirely new. Businesses that started out as closed have become more open over the past several decades. Robert Kahn noticed how the opening of the computer business made way for innovation in networking and personal computers. It also started up processes for standardization. Kahn[4]:

In the 1960s most of the business was in the hands of big established companies like IBM, Burroughs, Honeywell, and GE. It got more entrepreneurial when personal computers were introduced. Things changed in the late 1980s and 1990s when it became clear that computing was more of an open field rather than a closed field. When the Internet came along, it opened up even more.

In Henry Chesbrough's analysis, closed innovation is the old model. Closed innovation employs an implicit strategy of hiring the smartest technical people in an industry. It assumes that a company must itself develop its own new products and services and be the first company to get them to market. It assumes that the company that leads the industry in R&D spending will eventually lead the market. Finally, it assumes that a company should hold on to its intellectual property tightly to keep the competition from benefiting from the ideas.[5]

The strategies for closed innovation are subject to important failure modes. One is that highly experienced and skilled people are sometimes quite mobile. Mobility breaks the knowledge monopoly. Not only can employees join competitors; they can also form their own companies. In open innovation, new products and features may be developed outside of a company, and venture-capital firms can help the new companies develop new markets and new business models. Competition by startups crucially changes the cycle by providing the means to do experiments in new markets.

Closed-innovation models rely on captive research organizations to create the innovations that a company will need. In open-innovation strategies, the shift is from these one-to-one arrangements to many-to-many arrangements. In open-innovation companies get their technology from multiple sources. Open strategies for innovation seek efficiency through effective partnering.

Closed innovation is not all bad. This is the main model that was used by technology companies in much of the 20th century. It is a process by which companies fund investment in research and development, which enables breakthroughs, which leads to new products, which yields increased sales. This creates the ability to repeat the cycle.

Strategies and Limitations of Outsourcing   One of the case studies in Chesbrough's book *Open Innovation* contrasts the Xerox Star business model and the IBM business model around 1981. Both companies were marketing what might today be called a personal computer, but the contrast in their commercial successes with personal computers is striking. Open innovation won. Closed innovation lost. This sums it up, but the summary and conclusions turn out to be too simple when we consider the results over a few years.

The Xerox personal computer was a leading-edge high-performance computer with the ability to send, receive, and print high-quality documents. Although the IBM PC of today resembles the Xerox Star in its capabilities, the 1981 IBM PC did not. It was personal computing made affordable. Xerox built the entire Star system, from chips through software. including manufacturing, distribution, service, and financing. You could not really buy a single Xerox Star. You could buy a three-user system with a network facility and a shared laser printer for about $17,000. In contrast, the IBM PC cost about $3,000 and was marketed broadly to individuals and small businesses.

Among the factors in the success of the IBM computer were the brand name, the much lower price, and the use of third-party dealers for hardware and software. The product itself set the paradigm for open innovation. The processor came from Intel, the operating system from Microsoft, and the application software from third parties.

What worked for IBM in the open-innovation approach was that it powerfully exploited the strengths of its partners. IBM did not try to create or control all of the technology itself, and market forces quickly created competition among suppliers. IBM recognized that the product was a *personal computer* rather than a corporate computer.[6]

Although Xerox had been marketing the Alto and the Star for several years, it was disastrously slow to recognize and adopt more effective strategies. Furthermore, it did not powerfully exploit the strongest methods of the closed-innovation paradigm. For example, it did not aggressively protect its technology position with patents. It did not develop an effective licensing strategy, nor did it use patents to slow down the competition.

Without question, IBM's open-innovation approach to the personal computer dominated the market. However, it is interesting to remember that IBM's *execution* of open innovation did not keep IBM in the driver's seat. Today personal computers are more often referred to as "Wintel" computers than as "IBM computers." That is an acknowledgment that architectural control and market dominance now are driven by the Windows operating system from Microsoft and the processor chips from Intel. Lacking any proprietary advantage, IBM lost its hold on the PC market. Leadership passed to the companies with the key technologies.[7]

This outcome reveals some of the limitations of open innovation. When a sector turns completely to outsourcing and strong industry standards, commoditization begins. Commoditization brings broad competition, price wars, consolidation, and decreasing profit margins.

A second outcome of extreme outsourcing and standardization is resistance to change. Over an extended period, standardization is a barrier to invention. Once an entire industry gets organized around the detailed specifications and interfaces among standard parts, there is much inertia. A standard commodity resists radical improvements and change until it is replaced from below by another radical innovation.

Strategies for Leveraging Home-Grown Technology   Creating new knowledge and know-how is expensive. Sometimes companies create new methods or processes that have uses outside their own businesses. Companies try to protect their unused breakthroughs with patents, and try to reclaim value by enforcing patent rights and offering licenses. In many cases, however, the potential value of new intellectual property is difficult to evaluate and its usage is difficult to monitor. These uncertainties and the high costs of negotiation and legal arrangements have led to an inefficient market for intellectual property. These inefficiencies cry out for new strategies.

The second leg of strategies for open innovation is to leverage the strength of partners to get the most benefits from technologies that a corporation develops. In the old model, a company with a new technology would try to leverage it to gain market dominance. In the extreme form, it would patent it securely and try to go it alone in developing and marketing products.

There are several fundamental problems with this approach. Few products depend on just one technology. If a technology is useful but artificially expensive, creative minds will focus on finding alternative technologies and getting around patents. The resources for developing and exploiting a new technology often don't reside within a single corporation. Furthermore, if a new technology is outside the mainstream business of an organization, it will usually find that it lacks the skills, the will, and the resources to get the most from the technology.

There are several strategic elements in open innovation. First, companies can be active buyers and sellers of intellectual property. With specialized licensing partners, they can set up active programs for licensing technology beyond their own businesses. IBM and Texas Instruments have found major business opportunities for licensing their technologies, such as IBM's data-storage patents.

For companies considering a transition from closed to open innovation, Chesbrough advocates a process of taking stock: Where have the important ideas for your industry come from over the past few years? What role have startups played? How do the bets placed by venture capitalists compare to the bets made by your company?

The focus on venture capitalists and startups reflects both the faster cycle times of small companies and the creativity that goes into creating new product categories and new markets. Venture capitalists gain information from the "deal stream" of proposals. They are in a position to see synergistic combinations of ideas and technologies that are emerging at the same time. This gives them a privileged position and knowledge that is generally not available to closed-innovation companies.

Strategies for open innovation embrace these ideas about market creation. Rather than envying or just criticizing how venture-capital firms do it differently, these strategies enable companies to explore partnerships and try to learn from their example. The goals are to create synergistic opportunities and to understand blind spots in a company's strategies in order to create more effective road maps for future business.

What Startups Really Do   Open innovation is about creating efficient markets. In the 1990s, there was great experimentation in creating markets, especially in the dotcom movement. To the extent that the ideas of open innovation imitate venture-capital practices, it is important to pay attention not only to what worked but also to what did not.

The language of the "new-economy" and "old-economy" companies sounds quaint and naive after the dotcom crash. In the 1990s many people expected the new-economy companies to take over and the old-economy companies to fade away. That did not happen. When the bubble burst, the opposite did not happen either. A lot of new-economy companies died off; however, some did not, and many so-called old-economy companies also died off. We are left asking "What did the dotcoms companies teach us?" Reflecting on this question, John Seely

Brown distinguishes between experiments in technology and experiments in markets[8]:

I think of a dotcom as *innovation in markets*, not innovation in technology. The technology deployed by the dotcoms was by and large already there. What they were doing was imagining and trying to develop new kinds of markets using the Internet.[9] The construction of brand new markets has unprecedented amounts of uncertainty and risk to it and the outcome is determined solely by what really does catch on. When new markets catch on, social practices start to change.

The venture capitalists don't know how to think about that very well. Their mantra is "Will the dogs eat the dog food?"[10] What we are really talking about is not whether the dogs will eat the dog food, but what will end up changing social practices.

Now, 8 years after the dotcoms appeared, Amazon, Google, and Yahoo! have fundamentally changed our social practices for how we acquire information and how we play with ideas. It's just shocking to think what it was like 8 years ago—before we could search the web. The resources at your fingertips today are unprecedented if you are part of the digital culture.

Brown's analysis pinpoints crucial lessons of the dotcom period. The dotcoms did not create new technologies. From a technology perspective, most of what they did was akin to the incremental improvements by product-development groups late in the development cycle. They were pioneering in experimentation with new markets. In short, they explored "What is needed?" but did little to explore "What is possible?"

The venture-capital and startup culture in Silicon Valley and other places has been established for several decades. Venture capitalists are concentrated near Stanford University in Palo Alto and on Sand Hill Road in Menlo Park. They create new companies. Their role has been not only in aggregating funds for investment, but also in recruiting and organizing management teams and managing the early choices for fledgling new companies.[11]

The venture-capital culture has become a significant factor in the innovation ecology, creating options outside the base of established companies with their research labs and universities. This community has been active not only in the rise and fall of the "dotcom" companies but also in the computer companies in Silicon Valley and in the emerging biotech sector. In a sense, the approach of the venture capitalists is a model for the strategies of open innovation.

Dotcom Boom and Bust   The economic conditions of the new century began with the collapse of the economic bubble of the 1990s. Bettie

Steiger is a business consultant who has worked in Washington, D.C. and in Silicon Valley. She characterizes the 1990s as fostering unrealistic expectations for sustainable profits over the short term[12]:

During the 1990s all companies were expected to have at least 15 percent growth in revenues every year. That expectation was translated to the stock market and to the financial analyst market. A company that did not perform was down rated.

Management had more difficulty defending and investing in long-term projects because of the demand for financial performance. Unfortunately, you can't have sustainable growth without reinvestment. At a deeper level, there has to be a turn around in the thinking of stockholders demanding quick profits. I am not excusing the creative accounting by management, but there was a pressure to perform that they could not meet. When they could not show 15 to 18 percent growth, they started getting creative such as booking future sales as real revenue. And they *decreased* spending on innovation so that they could show revenue rather then cost in the short term.

One of the reasons that the dotcoms went under is that venture capitalists set unrealistic expectations. They wanted to double their investment in 3 years. In the 1990s venture capitalists were seen as the white knights and great creators of value for the country and for people.[13] Unfortunately, they were mostly creating high expectations out of vapor. There was no substance behind it.

What gave entrepreneurs or pseudo-entrepreneurs the bravado to create these companies that had no value? People created dotcom companies out of vapor and we invested in vapor. But why did we invest in vapor? All of a sudden the market was going up and the standard stocks no longer held what I would call charisma for the investor. People were looking for the short hit. The whole idea of investment for long-term gain was lost in the hype of the market.

People believed that the Internet was a "winner take all" market and that the first company to market would get the lion's share of the business. The future value of the company would be determined by the race to the Internet market, rather than by any technological advantage or sustainable business advantage. More often than not, they were racing to a market that didn't materialize.

During the 1990s companies were sometimes classified into "new-economy" and "old-economy" companies. New-economy or Internet companies asked "Who needs more innovation when our evaluations are already growing so fast? New innovations would just slow us down." Old-economy companies asked "Who can afford research when we have to trim costs to make our profit numbers?"

Several lessons from the dotcom period can provide guidance beyond the strategies of open innovation. Experimenting in new markets was a good idea, but the drive for quick profits was unrealistic and unsustainable. Creating markets for obtaining new technology is also good, and so is creating markets to exploit home-grown technology. However, these strategies—the core of open innovation—are not enough. The strategies of the dotcoms shortchange us in the end because they dampen interest in the long term. This brings us to the other half of the equation: new strategies for creating the breakthroughs that can solve new problems and create the potential for sustainable growth.

Open Invention

Open invention is the part of innovation that focuses on how breakthroughs get created. Nobody ever got a breakthrough from open innovation. In the extreme form of open innovation, everything is outsourced and products are low-margin commodities. This begins the end game for a technology. The next wave will be led by those who have breakthrough technologies.

Open invention tries to understand the future problems of future customers. It recognizes that two things are difficult: figuring out what problems the world *will need* solved (What is needed?) and creating technological solutions and breakthroughs to meet those needs (What is possible?). Thus, open invention searches both for what is needed and for what is possible. Strategies for open invention address that challenge.

Business and Its Future Customers It is not surprising that continuing along in a familiar direction does not always take companies where they need to go. Chuck Thacker, who now works for Microsoft, has been worried that new directions are needed for that company:

Now my worry is that the industry is busy making things that people don't actually want because we have saturated them with technology. This is a serious Microsoft worry. The sales of personal computers are not going up. They are flattening out and people are beginning to care about different things. When I started at Microsoft . . . , I had a discussion with Nathan Myhrvold about the value of making products more reliable. He said "No one will pay you a nickel for it. Ship frequently and don't worry too much about the bug rates. They will get fixed in the next release and people are fine with that." I didn't believe it because I came out of the UNIX[14] culture. But he was absolutely right at that time. People thought it was okay to reboot your machine five times a day. Rebooting didn't take all that long, and the computer would do such wonderful things for you that the delay was an acceptable price to pay. The attitude has shifted now.

Something else I am concerned about is that you hear more people acting negatively toward technology. When I was young, people were quite worried that computers were going to do great damage to the world and be completely dehumanizing. There was George Orwell's *1984* scenario. That did not happen and actually computers have been quite liberating for people, particularly personal computers. But now we have a couple of generations of people who don't know any different and computers are not making things better for them in the way that they made things better for those previous generations. So they are not reacting as positively. Most people are not technophiles. They just want to get on with their lives.

Thacker's thoughts reflect the state of the computer industry, which seems to be at the top of its S-curve. Assumptions that serve a company well during one part of the S-curve can create hazards in the next part. For example, assumptions that a company holds about what kind of technology is needed during the high-growth part of the technology-adoption curve may leave the company stranded when the market saturates.

A dictum in most companies during their profitable and high-growth phases is to listen to the customer. Companies develop methods for tracking customers' requirements carefully so that their products are well positioned in a market and have the potential to dominate the competition.

But how do you listen to future customers? Today's customers can sometimes imagine things that they would like. They have more difficulty understanding what things could be possible with radical new technology. Inventors can sometimes create technologies that nobody needs or solve problems that nobody needs solved. The art of inventing the future is about more than listening. It also involves imagining and getting around or through obstacles. It is about new directions and breakthroughs.

Directions to the Next New Thing When we ask for a solution to a very hard problem, we may have to wait a long time before a solution can be found. Scientists and others start working in areas of high potential before specific applications are understood. Nations or companies with a head start have an advantage. Researchers hedge their bets by doing work in broad areas on the expectation that it will have payoffs in *some* application.

At the time of this writing, there is much discussion in the research community about "NBIC" (Nano-Bio-Info-Cogno[15]), a convergence of four major areas of science and technology. From his position at the Institute for the Future in Menlo Park, Paul Saffo looks at technology trends arising at the intersections of cutting-edge research in information technology, materials science, biotechnology, and energy[16]:

In our technology forecasts, we look at the spaces where fields are merging or connecting. The four big quadrants are information technology, energy, materials, and biotechnology. Basically, information technology is mature. The field has been innovating for a couple of decades, but it still has many surprises. Biotechnology is the next big thing.

If you look at the last 100 years of innovation, there is a pattern. About once every 30 years a new science discipline turns into technology and becomes the platform for entrepreneurial activities. At the turn of the last century, the big science was chemistry. The science gods were chemists and little kids got Gilbert chemistry sets soon thereafter. World War I was the chemist's war.

The middle third of the century was shaped by physics and physicists. The advances in physics fascinated everyone. World War II was ended by the physicists, with developments in radar, atomic energy, and so on. The final third of the century was driven by information technology, which has transformed information sharing, logistics, and communications.

Each one of these revolutions builds on the earlier revolution. It was the experimental work of chemists like Madame Curie who helped connect with theoretical contributions of Einstein that turned physics into an industry at the middle of the century. And it was three physicists, Brattain, Bardeen, and Shockley, who were responsible for triggering the electronics revolution with the transistor. Right now we are on the cusp of a shift from information technology to biotechnology, with biotechnology becoming the main source of innovation and entrepreneurship. This doesn't mean that innovation in information technology is over, but I actually think that the big innovations in information technology are going to come from biological inspiration.

Even though everyone has their attention on biotechnology for genes and proteins, the short-term impact is biology inspiring how we think about information technology. We're going to have an explosion in the amount of information to be tracked but our traditional data systems of clients and servers just won't do the job. We need some radical new ideas about computational architecture and the like and I think that means that we're going to borrow the ideas about those structures from biology.

Material science interacts with information science in the space of things like MEMS[17] and arguably molecular computing. When you look at the field of energy it gets interesting because everything interacts with energy. You've got information technology for controls, you've got material science giving you new kinds of photovoltaics,[18] and you've got biotechnology interacting with energy to create engineered biosources and the like.

As a science becomes industrialized, it becomes more difficult for an individual researcher to build a significant reputation on individual work on a small scale. This has long been true in physics, where expensive facilities and teams to run equipment have been the norm for many decades. The shift toward big science is now dominating the leading edge in biotechnology. This shift has implications for research labs, and also for federal funding of research. Peer-reviewed, investigator-initiated research becomes less relevant as a science becomes more industrialized. As science gets bigger, leading-edge research requires institutions with critical mass of researchers and equipment.

University Research Institutions When research centers create a focused context for combining the efforts of scientists and engineers, the results can be magical. Scientists aim to increase knowledge. They publish papers that fill gaps in human knowledge. Engineers focus on building things and solving problems. In effect, engineers "publish" by building artifacts. John Hennessy sees the creation of new centers merging biotechnology and engineering as opening new doors for leading-edge research in the next decade[19]:

Engineering has a major role in most if not all of the multi-disciplinary research institutes that are springing up, including the newest ones involving engineering and biology. Engineers are problem solvers, and they are also risk takers because they often set out with a concept and work to demonstrate that that concept or insight can solve a problem.

I think that one of the great visions that everybody has in medicine is translating research from bench to bedside. This means being able to take the incredible understanding that we now have about the human body and biological systems starting down at the very basic DNA protein level analysis and moving up through the organs. We now understand how kidneys really work, how livers function, and we are applying that understanding to the care of the patient.

Succeeding at this requires *translational research*, that is, "engineering." Engineers are all translational researchers. They are people who take basic scientific understanding and apply it to real world problems. Engineering is a critical player as we move forward with Bio-X[20] at the Clark Center. Twenty years from now you are going to see large numbers of engineers who when you ask them about their core science, they will think about biology as much as they will think about physics and chemistry and the traditional disciplines on which engineering is built. This is going to lead to lots of new innovations because such people will approach problems from an engineering perspective. How do we go from point A to point B? How do we go from basic scientific understanding to good patient outcomes?

To bootstrap project creation, we created a system to provide seed funding for research groups engaged in multi-disciplinary work. The rule is simple. You have to have a multi-disciplinary team and you have to have a plan to build something bigger if the initial research succeeds and the initial collaboration succeeds.

When you seed things in a research university, the first thing that the faculty naturally thinks about is having a graduate seminar. After that they start collaborative research projects. Then they think about what the classroom environment should be and what we should be teaching at the graduate level. Eventually they think about the undergraduate level.

Choosing problems is crucial. My colleague Dick Zare in chemistry once said to me that there is lots of good research going on but there is much less *important research*. You frame this question differently in the sciences than in engineering. In the natural sciences the primary focus is increasing understanding, but obviously there are areas where new understanding could represent a great leap forward and there are other areas where it would be more incremental. In incremental research you add to the fact base and build on the foundation, but that is not the same as building a whole new foundation. I think that it is easier to assess this distinction in engineering because there you are trained to look at the scale of a problem and what the potential impact could be.

Scientists try to add knowledge. Obviously there is some notion in science about which knowledge could be more important or might have more impact. But because the time scale is longer in science, it is harder to know what impact it might eventually have. In the sciences you need a longer time frame to evaluate the work.

Hennessy's focus on the combination of science and engineering emphasizes the two parts of radical research. The engineering part focuses on solving an important problem. The science part, which is multi-disciplinary, creates the new knowledge that is needed to surmount obstacles. The culture of radical research uses both parts in following a problem to its root.

At times, the opening of new areas means that research needs to move out of universities and into new companies. Joshua Lederberg cites an example of this from his experience as president of Rockefeller University[21]:

[Since the mid 1990s] there has been a dramatic shift in the opportunities offered by the private sector at every level, and it goes far beyond compensation. It also includes investment in equipment for developing applications. One of our brilliant people at Rockefeller University decided he would do better working for Structural Genomics in San Diego than maintaining a chair here. I could see why because he probably had a budget that was 10 to 20 times larger than what he could get at the university. We are seeing more and more of that kind of competition.

There is a big overlap between his research at the university and his work at structural genomics. The scientific content is hardly distinguishable between what he is doing now, except that he is just doing a lot more of it and doing it in ways that might generate a revenue stream. He is industrializing protein structure determination and can crank out 1,000 structures a year instead of two or three.

In order to get the funding to make that feasible, he has to have a business model where by he can sell the results to the pharmaceutical industry. There are some compromises involved but the overall productivity is greatly enhanced.

Structure determination is central to pharmaceutical development these days. There is no way in the world that that research is going to be let alone. This issue is a challenge for the leadership of the universities: How much of this kind of research should we still be doing?

In Lederberg's example, determining the structure of a protein is shifting from being a difficult and long-term research project toward being a more routine technical step in drug development. This migration of an area of research into industry reflects a shift toward big science in scale and funding. An implication for the innovation ecology is that the area between big multi-disciplinary institutions such as the Clark Center at Stanford or Rockefeller University and corporate research laboratories is becoming more blurred. Finding effective and equitable ways to coordinate research, to do joint projects, and to avoid getting mired in negotiations or disputes about ownership of intellectual property are major challenges.

Institutions for Open Invention   Research organizations such as SRI International, Bell Laboratories, and PARC are moving toward open models in which they work with multiple corporate partners as well as with the U.S. government. Having multiple long-term partners is intended to reduce the cost of research through cost sharing on topics of common interest. This facilitates access to multiple manufacturing organizations and marketing channels.

This arrangement has several potential advantages over multi-disciplinary academic institutions. By participating as owners or long-term sponsors of research, corporate partners have more of a sense of commitment and participation. The purpose of the organization is simpler, since it is not mixed with an educational mission. Its main goal is to develop innovations, especially breakthroughs. When multiple partners are in different businesses, there is less risk created by sharing information and developing common ground. Each company has a different field of use and rights to inventions that are created. Access to and rights to intellectual property are predictable and pre-negotiated.

Several issues arise in creating and managing an open-research institution. Perhaps the primary issue is the challenge of serving multiple masters. Failure modes could arise if disagreements among the partners leads to fragmentation of the research organization. That would defeat the desired synergies, where partners combine resources and the research for one partner benefits the others as well. Addressing this requires the selection of compatible partners. The first partners are deeply involved in the selection and board approval of the next ones. Since synergy requires that the center operate in an open fashion, that intention should be reflected in the governance structure of the organization in order to prevent partitioning along partner interests.

A research company can also develop some inventions using government sponsorship. The aggregation of funds from multiple sponsors shares the costs and creates flexibility. It also creates a unique and strong basis for forays into new areas where breakthroughs and radical innovations are most likely to occur. This approach provides a stable basis for developing inventions and attracting venture funding as needed in the cases where the innovations fit none of the current owners.

Practical approaches to open invention should address both the costs and the benefits of funding breakthrough research and creating intellectual capital. They can help companies shift to strategies more effective than a high-risk/high-benefit game. For example, companies can reduce the costs and risks of developing new technologies through partnerships that share the costs of strategic common research. This strategy makes several useful tradeoffs. It gives up exclusive ownership of an invention for shared ownership. This trade has little downside when the sharing is among companies that use the same technology in different markets. It reduces the cost of the research by sharing its costs and sometimes by pooling complementary assets.

In contrast with universities, which are increasingly claiming intellectual property and negotiating high licensing fees, research institutions establish long-term relationships with their corporate partners for repeated engagements on many projects. In repeat engagements the terms of engagement are established up front, avoiding high negotiation costs about value or markets. From the perspective of each company, this leads to lower research costs and potentially access to complementary assets. It makes it possible for them to explore more options in innovation at a lower investment.

Research organizations can also work with venture capitalists to create spinoff companies from appropriate projects when the applications are outside the business interests of the partners. Partners have options to participate in the funding and success of the new startups.

Both changes—multiple corporate sponsors and venture participation—make it easier for researchers to see opportunities for positive impacts from their research. David Fork is representative of researchers who have been intrigued by the power of spinning innovations out of research centers when they don't fit the marketing capabilities of a single sponsoring organization[22]:

The power of a startup comes from the incentives it provides for people to put in long hours and get things to work. The founders focus all of their attention and bet their livelihood on its success. You can spin off the technology and use the venture-capital system to fund the work. There is a lot of capital investment and a great deal of capital development needed after the "Aha" insight. It's really hard to scale an effort in this way inside a research center.

Research often leads to applications not anticipated by the inventors or a single sponsor. The new applications of technology have potential value, but a corporate sponsor may not be equipped to commercialize them. Researchers and management look to spinoffs or licensing approaches that can tap into the opportunities created by unplanned applications of inventions. From the perspective of researchers who want to see their inventions get out into the world, spinoffs can get around internal roadblocks when the sponsoring organization is ineffective or uninterested in the technology.

A company making a decision about whether to partner with a research institution must consider several crucial questions:

Are we already getting the technology and breakthroughs we need from our own corporate research groups?

Does the research institution have fundamental capabilities or a culture of innovation that is missing from our own corporate research groups?

Does the research institution have a track record in creating breakthroughs?

Are the businesses of the other partners of the research institute complementary or competitive to our own business?

Do the partners' businesses have common technology needs, so that they can share the expenses of shared, strategic research?

There is more than one way to estimate the potential value of a partnership with a research institution. The technology and business parts of a company will look for synergies. How can we combine the capabilities of our company with the capabilities and assets of the research institution to create value? If the proposal is to invest as an owner, there is another perspective that will be considered by the "investment" group of the company. Corporate investment groups are used to investing in independent businesses, judging their value according to revenue, expenses, and expected market growth. These investment methods were not designed for comprehending or evaluating research institutions. The revenue of a research institution comes from its sponsored research projects, from government grants, and from licensing rather than from the marketing of a tangible product. There can also be occasional growth in value from spinoffs over time. Applying the usual kind of investment logic to a research partnership fails to account for the intent for the company and research institution to work together. For example, the investment group may balk when it realizes that part of the "revenue" of the research institution will come from annual participation fees that partners must pay. Failure to anticipate and address these issues can make it very difficult for a company to understand the value proposition or to decide to work with a research institution as a partner.

Open-research organizations can use their connections to have powerful conversations with multiple partners very early in the invention cycle. These conversations are more than the traditional "conversations with customers" associated with closed innovation. Researchers and people from different parts of the sponsoring company come together to explore the challenges of a market, perceptions of the hardest problems, trends, and emerging needs. They bring multiple disciplines for looking at a problem. By developing programs that temporarily assign company specialists to the research organization, in-depth bridges and long-term connections can be built.

Finding the Right Level for Risks

Innovation involves taking risks because it involves working with the unknown. According to Paul Saffo, innovation ecology needs the right level of risks and rewards:

There's an old saying that "Good judgment comes from experience. Experience comes from the exercise of bad judgment." If we don't create ways for people to safely fail, we've got problems. Here's the difficulty as I see it. It's not just "safely fail."

There's a Goldilocks zone[23] where innovation happens. In this zone it is neither so perilous that you don't dare take a chance, nor is it so soft and cuddly that you don't take innovation seriously. If you go below the Goldilocks zone, you can plod along in the same old way. There is absolutely no cost to failure and therefore no incentive to innovate. That's why most intrapreneurship[24] at corporations doesn't work because there are no consequences to messing up. On the other hand, if you are in an environment where it is very dangerous to fail, you are not going to innovate. To me the art of fostering innovation is to create those kinds of environments where it is safe for people to fail but they also feel the consequences of failure and are nudged back toward succeeding.

We need to think through levels of risk. Every institution from the smallest startup to the federal government has to ask: How do you create this Goldilocks zone? If you don't, the startup won't have the next idea that helps it a week from now. The government needs to incent the economy to be competitive. If the federal government doesn't get its policies and educational investments right, our children or grandchildren will discover that they are just bystanders in the world economy.

Taking the idea of a Goldilocks zone seriously involves more than institutional policies about benefits of taking risks and consequences for failures. There are also implications for institutional organization. The abundance of breakthroughs in the breakthrough zone is a consequence of the culture of innovation and the setting of corporate research labs. Because research labs are located midway between the concerns of development ("What is needed?") and the concerns of basic research ("What is possible?"), neither hurdle is impossibly high. They are positioned in a Goldilocks zone where more people can successfully jump both hurdles. Creating and sustaining such zones in the innovation ecology is a key factor for success in the new century.

Reflections

The boom-bust economy has biased public and investor thinking toward the short term, even though it has now become clear that the short-term growth of the 1990s was not sustainable. We now need to cultivate a long-term perspective and cultivate an attitude of custodianship.

Ultimately, handling the surprises of the future will require new innovations. To foster innovation, it helps to understand how innovation works.

The challenges are not easy. Strategies for open innovation address the challenges of acquiring and marketing technologies. Strategies for open invention address the challenges of anticipating future needs and creating breakthroughs to meet them. The companies that will lead the way will employ strategies of both open innovation and open invention.

Afterword

Mark Stefik

The Writing of this Book

In December 2000, research life at PARC was about to change. A boom in Silicon Valley was accelerating the creation of new businesses, but life inside PARC had not changed very much. The institution thought of itself as an *instrument* of change. It was where personal computers, the Ethernet, and laser printers had been invented. It had kicked off ubiquitous computing. It had spun off various companies deliberately in areas like laser technology and information visualization. PARC was famous for "unintended spinoffs" like Apple, Adobe, and 3COM. Xerox PARC was used to provoking change on the outside without changing too much on the inside.

Silicon Valley was consumed with Internet mania, and the world of research was trying to adjust. The economic bubble was in full force. Some people had left research to start companies. Everyone had friends who were becoming wealthy working at new start-ups. At the same time, a career in research seemed to be getting more precarious. Several research organizations in the Bay Area had closed, including Interval, Atari Labs, Apple's Advanced Technology Group, Schlumberger Labs, and Lucent Labs.

It became clear that Xerox PARC's parent company was in trouble. It was a few months before the NASDAQ would crash and many other technology companies would be in the same boat, but Xerox's stock price took a nose dive from over $65 a share to less than $10 a share. Xerox PARC was seen as an expensive institution in a shrinking company.

Xerox PARC began transitioning from being a corporate research center to becoming a research company—PARC (without the Xerox prefix). PARC needed to redefine its relationships and to develop a new financial basis for supporting its activities. Researchers and managers at PARC were concerned that the qualities that made the institution so productive and exciting might be lost in the transition.

I was appointed to direct one of the six laboratories at PARC. Together with the other laboratory managers and center director, I suddenly had a part in charting a new course for the institution. It quickly became clear that PARC and Xerox were in for a chaotic ride. I had many moments where I felt that I didn't deeply understand what to do.

Much of the PARC community was thinking about the changes ahead. There was concern not only that PARC needed to make changes in order to thrive in the future, but also that it should not accidentally undermine the culture that made the institution so innovative.

In academic circles, a time-honored way to learn about a topic is to teach a course in it or to write a book. Although this may seem strange and detached to people unfamiliar with the practice, the basic idea is sound. Writing can be a formal practice for reflecting deeply. When ideas are written down, they become available for examination, debate, and criticism. Writing and review lead to clarity about what the issues are.

This book was not written as a scholarly exercise, but rather as part of a determined and pragmatic effort to understand what was happening in industrial research and what needed to be done. Change was afoot. New economic realities surrounding invention and innovation were setting in, new policies were being tried, and PARC's transition was underway.

During this process, I continued my day job, participating in research projects and directing the Information Sciences and Technologies laboratory while PARC was in transition. The rhythm became research management and sometimes crisis by day, followed by writing and reflection at night or over a weekend as time from my primary job permitted. There were several crises and dark periods while the book was being written. The economy was tanking. Although PARC was being protected by Xerox, the budget was shrinking and PARC had to make painful choices in letting some good people go. It was by no means clear how things would turn out for PARC. Would the research center survive? Would it change radically? Could the economic conditions sustain a culture that was optimized for creating breakthroughs?

The writing and working activities informed each other as the issues that came up in the governance of PARC, the doing of research, mentoring with colleagues, or interviews paralleled themes arising in the writing of this book. Writing the book created some quiet space to reflect and to figure out what to do. The process involved not only sitting in the library but also sitting in the fire.

My wife joined me in the writing process. Trained in transpersonal psychology, Barbara has a knack for digging for deeper meanings. She has a habit of tugging at a thread until it emerges with clarity. She was a touchstone for the authenticity of the conclusions that were emerging. She started doing her own interviews for the book, identifying themes that I had overlooked, perhaps because I was too close to them.

By the third complete draft, we realized that the book would have been a lot easier to write if we had known from the beginning exactly what we wanted to say. It became abundantly clear to us that many aspects of research and innovation were changing fundamentally. PARC was a crucible of change.

Innovation and Its Study

We began by taking an inventory of the culture and methods of innovation at PARC. As our writing progressed, we quickly determined that limiting our investigation to PARC would be too confining. We were exploring how inventors work and how innovation happens—seeking insights into the challenges of the new century. PARC was an exemplar in the innovation ecology, but the story was a lot bigger than PARC. Most of our contacts were in the information-technology sector. We expanded our networking to include interviews at other leading institutions.

Innovation and its relation to prosperity and business are increasingly international concerns as globalization links together the economies of distant countries. The deliberate changes in the governance and economies of the European Union are just one part of trends much wider in scope. Brain drains and "reverse brain drains" are spreading both technological know-how and cross-cultural business ideas. At any moment, the eddies of international commerce have different effects at different places. Although we have made an effort as authors to frame the issues universally, our primary experience base is American and we write from the perspective that we know best.

As we broadened our interests in innovation, creativity, and the history of technology, we had to set limits on the scope of our investigation to keep the project manageable. We kept our focus on technological innovation, especially breakthrough and radical innovation. Much has been written about creativity in general and about product innovation in industrial engineering and design. For example, Tom Kelley's book *The Art of Innovation* gives a compelling account of creative work at IDEO, a leading industrial-design firm. IDEO is known for its creative and user-centered product-design practices. However, its focus on short-term projects generally precludes radical or "big idea" innovation, and the account leaves out both the preparations and practices necessary for such innovative work. Some of the most far-reaching and innovative technological advances have arisen from the dialogue between engineering and science, each feeding and enabling the other. We were particularly interested in inventions at the edges of disciplines and innovations that stretch the limits of knowledge as well as creating important new kinds of things.

Another powerful kind of innovation concerns business process re-engineering. For example, Womack, Jones, and Roos's book *The Machine That Changed the World* describes innovations and techniques that have changed automobile industry by reshaping how it handles supply chains, inventory, and quality control. This kind of innovation is largely concerned with business and manufacturing processes. Lean production has changed the way manufacturing is done as much as mass production did almost a century earlier. Currently, the extensive use of outsourcing is having a similar effect. This kind of innovation clearly leads to important breakthroughs. It was, however, outside the scope of our direct experience. We focused our attention on inventions that are driven by insights in science and engineering. Technologies seldom stand alone. Increasingly, adoption and success require that technologies work together even when they come from different companies. It has become increasingly important to have industry-wide mechanisms for establishing standards. In his article "The Social Life of Innovation," (*Communications of the ACM*, April 2004) Peter Denning argues that standards organizations are essential for the success of many modern innovations, including the Internet. A discussion of this matter, however, is beyond the focus of this book.

In selecting people to interview, we focused on repeat inventors and innovators, believing that they were the most likely to have repeatable methods. We sought active practitioners whose experiences would be fresh. As with any sustained and deep practice, there are some aspects of experience that are difficult to access without direct personal experience. After some deliberation, we decided to address some of these by including some stories drawn from our own personal experiences.

We decided not to concentrate on famous scientists, inventors and managers because we wanted to understand more deeply about working in the trenches of innovation. We spiraled out from PARC, following a network of leads. Along the way, we interviewed about thirty people. About half of them were inventors or scientists at PARC and half were from other institutions.

Since we were also looking for insights into research management, we interviewed people who managed research or had roles in setting science policy or managing the business, education, and funding sides of innovation. In our interviews, we asked for stories and examples rather than speculations or theories.

Once the interviews were transcribed, we had more than 500 pages of typed material and 200 pages of additional handwritten notes. Reading the interviews was like being at a large party of very bright people discussing many aspects of innovation. As we worked through the material, we began to discern common themes in what people were telling us and in how they went about their work. A reasonably coherent picture began to emerge. Our descriptions of what seems to matter began to converge.

We also looked at a wide range of studies and books about creativity, invention, lives of inventors, business cycles, science policy, history of technology, and so on. Innovation is a compelling and challenging topic, and we found parts of the road well lit. We found the trail markings of earlier travelers—especially Meg Graham, Clayton Christensen, and Henry Chesbrough—specially helpful, and we hope we have helped to map more of the trail. The sources that we found to be the most cogent and accessible are listed in the bibliography.

In the end, we cannot claim that we have covered everything that matters in innovation. We wanted to target themes and issues at the heart of innovation, more than to survey its variations. We wanted to identify and understand the key issues confronting innovation and research organizations. We wanted to understand what choices were possible and, to some extent, which issues were timeless and which were new in the evolving ecology of innovation. We hope that the ideas expressed here help to ignite a love for this kind of work in those who are seeking a career in innovation, and that generally the book leads to deeper understanding in a society that depends on its own creativity to help shape a better future.

Notes

Chapter 1

1. The Institute for the Future, located in Menlo Park, California, is an independent non-profit research organization with expertise in forecasting and in mapping critical technological, demographic, and business trends. It helps its clients to make sense of the complexity of these trends so as to understand the social and business implications.

2. Paul Saffo's comments are drawn from an interview recorded October 10, 2002.

3. Engelbart is credited with inspiring many people with his early demonstration of a computer system with a mouse, hypertext, and multiple windows.

4. The term "creative destruction" was coined by Joseph Schumpeter. In his 1942 book *Capitalism, Socialism, and Democracy,* he argued that capitalism is a method of economic change and that it can never be stationary. Spurts of innovation inevitably destroy established enterprises and create new ones. Schumpeter saw capitalism as creating new goods, new methods of production or transportation, new markets, and new forms of corporate organization. In terms of technology-adoption curves, creative destruction is the process by which the market for technology decays as another one starts up.

5. See Geoffrey Moore's *Crossing the Chasm* for a discussion of technology-adoption curves from a perspective of growing market share. Moore discusses marketing strategies and the creation of "gorillas" in new markets. In *The Innovator's Solution,* Clayton Christensen and Michael Raynor consider the business trajectory of companies through their period of rapid growth. During this period, companies focus on their most important customers and improve their products to meet the needs of these customers. After enough improvements, the product will exceed the performance that these customers can use. This is the point where the market wll saturate and competition changes as products become commodities. Innovation at this point may take the form of a low-cost disruptor, radically reducing the cost and profit margins for the product.

6. The CV-2000 was developed by a team led by Nobutoshi Kihara. It was roughly the size of an audio tape recorder. It was much less expensive than the models created for broadcast use. This was a reel-to-reel unit and was limited to

black-and-white recordings. Sales were in the hundreds. The first usable proto-type videocassette recorder was developed by Kihara and his team in 1968. The first videocassette recorder sold commercially for home use was the Philips Model 1500, sold in England in 1972.

7. By June 2002, NetFlix had about 98 percent of the online DVD rental market. Corporate revenues rose from about $30 million in the first quarter of 2002 to $56 million in the first quarter of 2003.

8. In the semiconductor industry, the cost of equipment (capitalization) has gone up with each generation. In the jargon of the information sector, Moore's Law refers to the doubling of computer speed and memory size every 18 months. This doubling of performance has relied on advances in chip manufacturing, primar-ily photolithography, and has resulted in dropping prices for computation. In 1995, the manufacturing cost was about 17 cents per million data bits in the 64 M (megabit) chip. The cost for the same chip fell to 7 cents by 1998 and to 3 cents by 2001. It is projected to drop to 0.2 cent by 2010. However, keeping up with the technology has also required increasingly expensive manufacturing facil-ities. In 1980, a factory for a 64 K (kilobit) chip cost about $100 million to build. In 1985, the chips were 256 K and a factory cost $200 million. The cost has gone up with each generation of memory chips. By 1993, the cost of a factory was about $700 million.

9. Gordon Moore is known for his prediction that progress in semiconductor manufacturing would lead essentially to a doubling of performance every 18 months. As this rule has reasonably predicted progress in computer speeds and memory sizes for so many years, it became known as Moore's Law.

10. Two separate inventors, unaware of each other's activities, invented almost identical integrated circuits at about the same time. Jack Kilby was working at Texas Instruments, and Robert Noyce was working at Fairchild Semiconductor Corporation. In 1959, both parties applied for patents.

11. In early 1974, in England, Frederick Sanger developed the basics of DNA sequencing. Almost simultaneously in the United States, Alan Maxam and Walter Gilbert created a somewhat different method. The Sanger method has proved much better in practice. Even so, the Sanger method was relatively slow and tedious and required the use of x rays to read off the sequence. The technology has continued to advance. Methods using fluorescent tags were developed later. By the late 1980s, Leroy Hood's laboratory at the California Institute of Tech-nology developed the first semi-automated machine for sequencing DNA. In 1991, at Celera, Craig Venter developed a "shotgunning" technique that breaks large sequences of DNA into many small fragments, runs them through DNA sequencers that can read hundreds of bases at a time, and then re-assembles the results into a map using computer matching.

12. Arthur Kornberg discovered DNA polymerase in the late 1950s. However, it was not until the late 1980s that Kary Mullis introduced the PCR method. The technology has advanced further with the advent of various kinds of "DNA chips," microscale devices for controlling and routing biologically active fluids. At the time of this writing, pipeline chips for PCR are just being reported that can thermally cycle small volumes of fluid in 100 milliseconds, allowing, for

example, 20 cycles of DNA amplification in under 5 minutes. In this way, the biotech industry has followed the usual pattern of initial creation in fundamental research at universities followed by technology development at industrial research laboratories followed by product development in product organizations.

13. Most of the research in biotechnology apply to the medical and agricultural sectors. Products in these sectors face not only the challenges of understanding the unknown but also a myriad of hurdles related to safety testing, efficacy testing, and occasionally moral concerns. Work on a new pharmaceutical drug, for example, can go on for years only to be dropped at the last stage if it runs into snags getting approval.

14. See pp. 3–4 of Graham and Pruitt, *R&D for Industry*.

15. According to *Science and Engineering Indicators—2002*, funding for R&D rose from $169.2 billion in 1994 to $265 billion in 2000 (p. 4-3). Adjusting for inflation, this is an increase of $71 billion in 1966 dollars. Among countries in the Organization for Economic Cooperation (OECD), the United States is the largest single performer of research and development. R&D expenditures amount to 44 percent of the estimated $518 billion 1998 OECD total (p. O-12). This figure, however, includes both research and development. For basic research, the government agencies that have historically funded long-term research and technology reduced their share of the contribution to about 50 percent of the total, down from about 70.5 percent in 1980. According to the figures from the National Science Board's *Science and Engineering Indicators 2002*, "this decline in the Federal share of basic research support does not reflect a decline in the actual amount of Federal support, which, in fact, grew 3.5 percent per year in real terms between 1980 and 2000. . . . Rather, it reflects a growing tendency for funding of basic research to come from other sectors." According to *IEEE Spectrum's* R&D survey, the top 100 companies have been increasing private R&D expenditures. In 2001 budgets rose by an average of 5.21 percent. Because these numbers include funding for both basic research and development, they don't indicate directly which investments lead to new technologies. Development expenses, requiring by far the lion's share of investment, focus mainly on incremental improvements to products rather than developing new technologies. About 71 percent of corporate R&D is spent on product development rather than research. Definitions and reporting varies between institutions, but a rough rule of thumb is that even in organizations that do only research and no development, only about twenty to thirty percent of research is "basic" research, whose applications are either unknown or are at least 5 years out. A deeper difficulty in interpreting the meaning of research investments is that not all investments in basic research have equal potential. Some fields are entering periods of rapid discovery and invention, whereas other fields are more mature and are less likely to lead to new breakthroughs. A large fraction of the private research budget in big companies is in areas where the companies are already doing big business, and where from a broader perspective, the biggest breakthroughs may have already been realized.

16. Bell Labs and Interval were probably the most reported cases. Bell Labs and Interval were on very different scales in time, size, and ambition. Bell Labs had

about 24,000 researchers in 1999. It was organized as a major investment in basic and applied research and achieved breakthroughs in science and technology over several decades. Interval was a comparatively small research organization planned as an experiment for one decade. It intended to translate applied research quickly to business value. Lacking a built-in client or a compelling business case for its projects, the experiment was stopped several years early.

17. John Hennessy's story is drawn from an interview recorded October 10, 2002.

18. Chuck Thacker's comments are drawn from an interview recorded September 6, 2002.

19. Sand Hill Road in Menlo Park is where many of the venture-capital firms are located.

20. Although in its narrow sense 'dilemma' means *binary choice*, we follow Clayton Christensen's use of the word in his book *The Innovator's Dilemma*. The intended meaning is *conflicted choice*, especially where the conflicts tend to lead to bad decisions.

21. Chapter 11 analyzes cultures of innovation in terms of patron models and client models. Patron models are more tuned to breakthrough research, and client models are more tuned to incremental research.

22. John Hennessy's story is based on an interview recorded October 10, 2002. Hennessy founded the computer company MIPS.

23. See Clayton Christensen's book *The Innovator's Dilemma* for an extensive discussion of this issue.

24. See Stanley Feder, "Overcoming 'Mindsets': What Corporations Can Learn from Government Intelligence Failures."

25. Clayton Christensen described case studies of this dilemma in *The Innovator's Dilemma*.

26. Moore's position on research is articulated in his article in Rosenbloom and Spencer, *Engines of Innovation*. Curiously, his own success at Intel resulted after he, Noyce, and others left Fairchild Research to found the successful semiconductor company.

27. Over the years, PARC was "credited" as a "national resource"—much to the chagrin of Xerox outside of its public-relations department.

28. When universities set up multi-disciplinary institutions to attack major problems, they often face additional dilemmas. To attract funding to these institutions, they often solicit corporate sponsorship. When the sponsorship involves new entrepreneurial companies, the faculty themselves often have important roles in those companies. Derek Bok, past president of Harvard, has spoken about the institutional hazards of such arrangements. When students are encouraged to pursue a particular line of research, the question arises as to whether this is to further a commercial end of the sponsoring companies or the pursuit of knowledge relevant to a thesis. The dual purpose of these institutions, therefore, risks undermining the core educational mission of the university and the deep trust that is required between the students and the faculty. See Bok, *Universities in the Marketplace*.

29. The material in this section—especially the framing of the first five research eras in the United States—is drawn from the work of Margaret Graham in a joint forum talk given at PARC by Margaret Graham and Mark Stefik in 1991. Some of the analysis draws on Graham's article "Industrial Research in the Age of Big Science."

30. As part of the Omnibus Trade and Competitiveness Act, the National Bureau of Standards changed its name in 1988 to National Institute of Standards and Technology (NIST).

31. For several metrics of domestic and international investments in R&D, see National Science Board, *Science and Engineering Indicators—2002*.

32. According to the *Science and Engineering Indicators 2002*, "little venture capital is disbursed as seed money (high-risk funds to underwrite proof-of-concept activities or early product development). Seed funding never exceeded 6 percent of total venture-capital disbursements in the past two decades and, more typically, ranged from 2 to 4 percent."

33. The approach of following a problem to its root and working in the white space between disciplines is a part of the research culture at PARC and has been previously articulated by John Seely Brown. See the bibliography.

34. See Chesbrough, *Open Innovation*.

35. Chesbrough's studies of commercialization experiences at Xerox PARC were a primary source for his book.

Chapter 2

1. In some innovations, the two questions are addressed by different people at different stages in the process. For example, the "What is possible?" question may be more the focus of a researcher or inventor, and the "What is needed?" question may be more the focus of business development and marketing. Such a division of labor can be problematic in many ways. In general we have observed the early interplay of the two questions tends to be greater in organizations and individuals who have been involved with inventions that have been taken all the way to product and practical use.

2. IEEE: Institute for Electrical and Electronic Engineers.

3. Heterostructures are structures that combine two or more different types of material. Discontinuities of the material properties at the junction between them can provide enhanced optical or electronic performance.

4. Quoted from the *Spectrum* article.

5. ISO 9000 certification quality control programs are elaborate steps and checklists that are used to govern processes for quality control in products. Six Sigma is an example of a formal process used in product development to raise quality. These processes are designed more for routine and large-scale engineering projects than for radical, ground-breaking research. Kroemer's comments are representative of the dislike of these processes by creative people who have had them inappropriately forced on their organizations.

6. Ted Selker's story is based on an interview conducted April 18, 2001.

7. For an account of the challenges and adventures of this project, see Hiltzik, *Dealers of Lightning*.

8. Chuck Thacker's story is drawn from an interview recorded September 6, 2002. Gary Starkweather, who formulated the goal of creating a laser printer, came to PARC to work on the project after a technology group in another part of Xerox refused to support the research.

9. Ethernet is the local area computer network. Alto was the name of the personal computer built at PARC. RCG was an acronym for Research Character Generator; SLOT was an acronym for Scanning Laser Output Terminal.

10. The acronym dpi stands for dots per inch. In contrast to type-chain printers and typewriters, which used the impact of a shaped metal stamp to imprint a typeface, laser and inkjet printers build up letters by printing individual dots on a page. The greater the density of dots, the more precisely the fonts can be formed.

11. For several years, Thacker did research at the Digital Equipment Corporation.

12. John Seely Brown's comments are drawn from an interview recorded September 17, 2002.

13. DocuTech is a high-end printing and copying product made by Xerox. PARC developed solid-state lasers to improve its performance in laser printing.

14. SDL is Spectra Diode Labs, a Xerox PARC spinoff that was later merged with JDS Uniphase.

15. The inventors say that CUDA is not an acronym for anything and give no account of the meaning of the word. The project was later renamed ubitext.

16. Henry Baird's story is based on two interviews, one conducted by Mark Stefik on March 15, 2001 and one by Barbara Stefik on June 27, 2001.

17. HTML (HyperText Markup Language) is the language used in describing web pages. It contains the text of the web page as well as information about formatting and imaging information.

18. Maribeth Back's story is based on an interview conducted by Mark Stefik on April 12, 2001 and one by Barbara Stefik on May 11, 2001.

19. New knowledge included knowledge about how to create scalable, flexible fonts and how to use lasers in printing. In the larger context, new knowledge included knowledge about graphical user interfaces, WYSIWYG editors, the design of communication protocols for the Ethernet, and many other things.

Chapter 3

1. John Seely Brown has articulated the metaphor of seeing differently at PARC for many years and it has become deeply embedded in its research culture. In his book *Seeing Differently*, Brown explores new strategies for business innovation. His paper of the same title explores the role of multi-disciplinary approaches to research.

2. The four approaches were suggested to us by Joshua Lederberg in a personal communication dated October 12, 2002. He carried out an analysis of his own research and that of close colleagues. All four approaches were well represented, and most of the research was dominated by one approach. For need-driven invention, Lederberg cites his need to isolate auxotrophic mutants (bacteria requiring specific substances for growth and metabolism) as inspiring his invention of a penicillin method and also his invention of replica plating. For the data-driven approach, he notes that anomalous data led to the discovery of virus-mediated transduction (transfer of genetic material from one bacterial cell to another) and also the discovery of plasmids and lysogeny (the fusion of nucleic acid of a bacteriophage with that of a host bacterium). For the method-driven approach, he cites the design of instrumentation for exobiology (used in a Mars mission) and a use of replica plating to prove pre-adaptive occurrence of mutants. For the theory-driven approach, he cites his own review of the natural history of bacteria, which persuaded him that sex (the exchange of genes) was a likely process. This led to his Nobel Prize. Lederberg notes that sex in bacteria also involved a need element and a method element. To understand the implications of the first intimation that genes are encoded in DNA (Avery et al. 1944) required better insight into whether bacteria had genes. Also, the method of nutritional selection was used to pick out rare genotypes from mixed populations. As our conversations with Lederberg pursued and refined the understanding of the four approaches and related them to famous scientists and inventors, Lederberg exclaimed "How delightful to find this deep connection between Einstein, Edison, and others who have inspired me!"

3. Source: "What Life Means to Einstein: An Interview by George Sylvester Viereck," *Saturday Evening Post*, October 26, 1929.

4. Mark Yim's story is based on an interview recorded February 7, 2003.

5. The modular robotics research at PARC was funded in part by the Defense Advanced Research Projects Agency.

6. The book was Card, Moran, and Newell, *The Psychology of Human-Computer Interaction*.

7. Ted Selker's story is based on an interview conducted April 18, 2001.

8. GOMS stands for Goals, Operators, Methods, and Selection rules. This is roughly a time-and-motion analysis for a cognitive task.

9. Card's story is based on an interview with Mark Stefik on July 6, 2001 and one with Barbara Stefik on August 2, 2001.

10. Star was an early personal computer and office product by XSoft, a Xerox company.

11. Tiled windows are non-overlapping windows, arranged in a table or grid.

12. Locality of reference in window usage means that a user tends to use members of a group of windows for a period of time and then move to another group of windows.

13. A page is a unit of memory. For example, on some computers a page is 512 words of memory.

14. Herbert Simon was the Richard King Mellon University Professor of Computer Science and Psychology at Carnegie Mellon University. He received many honors in the course of his career. In 1975 he received the Turing Award for his work in Computer Science. In 1978 he received the Nobel Prize in Economic Sciences. He was instrumental in the development of the cognitive science group at CMU.

15. Allen Newell was the U. A. and Helen Whitaker Professor of Computer Science at Carnegie Mellon University. He had a bachelor's degree in physics from Stanford and a PhD in industrial administration from Carnegie Institute of Technology. He played a pivotal role in creating Carnegie Mellon's Department of Computer Science. His scientific contributions ranged from electrical engineering and computer science to cognitive psychology. Allen Newell was Herb Simon's student and later his colleague at Carnegie Mellon University. As a graduate student, Card worked with both of them.

16. The Alto was a personal computer developed by the Computer Science Laboratory at Xerox PARC, then led by Bob Taylor. The Alto introduced combinations of hardware (bit-mapped displays, mice, Ethernet, laser printers) and software (graphics user interfaces, visual text editors, drawing programs) that later became the mainstay concepts in the personal computer industry. Taylor was awarded the National Medal of Technology in 1999. See the bibliography for several books on the story of this invention and the drama of creating the industry.

17. Sheridon's story is based on two recorded interviews, one conducted by Mark Stefik on March 17, 2001 and one by Barbara Stefik on June 12, 2001.

18. Pattie Maes's story is based on two interviews, one conducted by Mark Stefik on June 28, 2001 and one by Barbara Stefik on September 7, 2001.

19. Sometimes an idea is discovered in more than one place or is rediscovered. In the case of collaborative filtering, both things happened. Pattie Maes's group at MIT and John Reidl's group at the University of Minnesota began exploring collaborative filtering at about the same time, exploring applications to music and movies and other things. About 10 years earlier, a system for collaborative filtering for email news had been built and reported on by PARC researchers David Goldberg, David Nichols, Brian Oki, and Douglas Terry. Their December 1992 paper "Using collaborative filtering to weave an information tapestry" was published in *Communications of the ACM* (35, no. 12, pp. 61–70). Reinvention and rediscovery are more likely when an invention is far ahead of its major application.

20. More generally, this mode includes new *methods* for doing experiments not done before, not just new kinds of instrumentation. The method should make new kinds of observation practical.

21. Figure 3.2 was suggested by Rich Gold, a designer and researcher who worked at Xerox PARC for many years. Gold was very reflective on the nature of design and its relation to the other creative professions.

22. Organizations made up mostly of designers often do very creative work and sometimes develop new inventions. However, they typically focus on inventions

made up from existing technologies, and they rarely come up with what would be called breakthrough inventions.

Chapter 4

1. From Richard S. Rosenbloom, "Sustaining American Innovation: Where Will Technology Come From," in the National Research Council's report on *Harnessing Science and Technology for America's Economic Future*.

2. This language arises from organizations that were divided into functional groups of fixed membership. People and their innovations do not travel across the metaphorical walls between groups.

3. The wire recorder was a recording device preceding the tape recorder. It recorded sounds by impressing a magnetic pattern on stainless steel wires. In general, the sound quality of wire recorders was inferior to that on tape recorders, and the recordings were difficult to edit.

4. Kihara's story is based on an exchange of email messages in November 2002 and the help of a translator provided by the Sony Corporation.

5. Computational linguists make computational models of language and are concerned both with knowledge and the complexity of algorithms that do natural language processing.

6. Ron Kaplan's story is based on two interviews, one by Mark Stefik on July 6, 2001 and one by Barbara Stefik on August 27, 2001.

7. The simplest linguistic models of language are the "finite-state" models. These models describe the recognition of linguistic elements such as sentences or words in terms of transitions between a finite collection of states. For example, the sequence of letters r-u-n might be encoded as three states: "start with an r," "r followed by u," and then "u followed by n." Branch points in the finite-state model would also recognize "r-u-n-s" but not "r-u-n-b".

8. Here 'space' refers to computer memory, that is, the space to store codes for letters and other symbols. The more symbols to be store, the more space is required.

9. It is often the case in the design of computer programs that there are trade-offs in the design between the amount of space that the program uses and the amount of time that it takes to perform a computation. For example, one way to trade space for time is to compute a lot of answers ahead of time and store them in a big table. Having a big table saves time because a program only needs to find an answer in the table, rather than performing a complex computation.

10. UMC means unit manufacturing cost. In estimating a price for a product, some companies multiply the UMC by a fixed percentage so that the price point (and profit) is sensitive to the UMC.

11. The 8086 was the Intel chip used in early IBM personal computers.

12. The TrackPoint is the eraser-like nub that appears between the G, H, and B keys on IBM ThinkPad laptop computers and many others. This section

considers the challenges in bringing the invention to market. See chapter 3 for an account of how it was invented in the first place.

13. Ted Selker's story is based on an interview conducted April 18, 2001.

14. Akihabara is a shopping neighborhood in Tokyo specializing in high-technology items. It is a popular stop for technology enthusiasts because items appear here long before they are shipped overseas.

15. Sorbothane is a viscoelastic material that isolates and dampens shocks and vibrations.

16. A pointing device is a device for selecting a point in a workspace. Examples of pointing devices include the mouse, the TrackPoint, trackballs, joysticks, and touch-sensitive screens.

17. Shneiderman's story is based on an interview recorded July 7, 2001.

18. The year 2002 was a very challenging year for Corning's business. In his written address to stockholders, James Houghton, chairman and chief executive officer, addressed how the company was responding to a very tough year in revenues. The most immediate cause of the difficulty was probably the downturn in the telecommunications industry and a reduction in demand for fiber optics cable. In the middle of the adversity, Houghton spoke to the company's sustained commitment to innovation. Although Corning had to eliminate over 19,000 jobs in 2001 and 2002, he said: "We are applying more than 10 percent of our revenues toward research. Some may question this high level of commitment in these times . . . but we simply will not back away from it. We have more than 1,000 scientists and researchers in our laboratories. They are at the heart of our innovation engine, and they're going to stay that way!" Corning has more than 150 years of investment in innovation, and has been depended on technologies in several periods of its history to pull it into new markets.

19. The 120-inch mirror became the main mirror for a telescope at the Lick Observatory on Mount Hamilton, near San Jose.

20. Corning Incorporated was originally known as Corning Glassworks. For a wonderful history of innovation at Corning, see Graham and Shuldiner, *Corning and the Craft of Innovation.*

Chapter 5

1. Paul F. Berliner, *Thinking in Jazz: The Infinite Art of Improvisation* (University of Chicago Press, 1994).

2. John Riedl's story is based on an interview recorded August 29, 2001.

3. Personal communication.

4. From an interview by Ernest C. Miller in 1980 for the American management Association. The text of the interview was sent to the authors in a personal communication.

5. Pattie Maes's story is based on two interviews, one conducted by Mark Stefik on June 28, 2001 and one by Barbara Stefik on September 7, 2001.

6. John Brown's story is based on an interview recorded September 17, 2002.

7. Larry Leifer's story is based on an interview recorded September 3, 2002.

8. IDEO, a product design company, was founded in 1978 by David Kelley and Dean Hovey and grew out of the Product Design program at Stanford. Tom Kelley's book *The Art of Innovation* talks about the culture and methods of IDEO.

9. John Hennessy's story is based on an interview recorded October 10, 2002.

10. Mark Stefik's story is based on an interview conducted by Barbara Stefik in August 2002.

11. AI: artificial intelligence.

12. Donald Kennedy, "Government Policies and the Cost of Doing Research," *Science* 227 (1985): 480–484.

13. Card's story is based on an interview conducted by Mark Stefik on July 6, 2001 and one by Barbara Stefik on August 2, 2001.

14. Marvin Minsky is one of the founders of the field of artificial intelligence.

15. Daniel Bobrow's story is based on an interview recorded September 3, 2002.

16. Junior Fellow at Harvard is a prestigious position. Another Fellow at that time was Noam Chomsky.

17. Papert had been affiliated with Piaget's institute.

18. John McCarthy was the inventor of Lisp, a programming language for processing lists and other symbolic structures. Lisp was used for much of the research in artificial intelligence because of its facility for modeling mental activity as representing and manipulating symbol structures.

19. Chuck Thacker's story is based on an interview recorded September 6, 2003.

20. "Cal": the University of California at Berkeley.

21. Before personal computers, people shared access to big computers. The computer would switch its attention from one user to the next so quickly that each person seemed to have the attention of the computer. Time sharing was a new idea in the late 1960s, enabling people to have interactive computing from terminals rather than "batch processing." In batch processing, typically users would put their instructions on decks of punched cards and submit "jobs" to a computer center for later pickup. The history of the invention of the personal computer is described in Mitch Waldrop's book *The Dream Machine*.

22. Ron Kaplan's story is based on two interviews, one by Mark Stefik on July 6, 2001 and one by Barbara Stefik on August 27, 2001.

23. David Goldberg's story is based on an interview conducted by Mark Stefik on March 29, 2001 and one conducted by Barbara Stefik on April 2, 2001.

24. Diana Smetters's story is based on an interview recorded April 12, 2001.

25. According to *Science and Engineering Indicators—2002*, more than 2.6 million students earned a first university degree worldwide in science or

engineering in 1999. More than 1.1 million of these degrees were earned by Asian students at Asian universities. Next in production were the European countries, with about 280,000 degrees in engineering and nearly a million science and engineering degrees overall. The United States had about 500,000 science and engineering degrees overall, with the smallest fraction in engineering.

26. Peter Medawar's paper was published in *Saturday Review* (47, August 1, 1964, pp. 42–43).

27. Several books have been written about discovery in mathematics, including Jacques Hadamard's *The Psychology of Invention in the Mathematical Field*, G. H. Hardy's *A Mathematician's Apology*, and Imre Lakatos's *Proofs and Refutations*.

Chapter 6

1. Many variations of this quote have been cited. Friends of Alan Kay at PARC remember the quote starting out at around "10 IQ points." Recently, versions with numbers as high as "80 IQ points" have appeared. In any case, Kay's point is that a point of view can provide a real advantage for having insights and reasoning by analogy.

2. Riedl's story is based on an interview recorded August 29, 2001.

3. EMACS is a programmable text-based editing system.

4. Alan Turing was a mathematician who worked out many deep problems about the limits of computation.

5. Ramanujan was a mathematician famous both for his brilliance and for being self-taught at an early age.

6. Pattie Maes's story is based on two interviews, one conducted by Mark Stefik on June 28, 2001 and one by Barbara Stefik on September 7, 2001.

7. Ethology: the study of animal behavior.

8. Shneiderman's story is based on an interview taken on July 7, 2001.

9. Mullis documents his discovery in his book *Dancing Naked in the Mind Field*, published in 2000. The discovery was reported in Mullis, K. B., et al., *Cold Spring Harbor Symposia on Quantitative Biology* 51 (1986): 263–273.

10. He says that the insight occurred at mile marker 46.58 on Highway 128.

11. Ron Kaplan's story is based on two interviews, one by Mark Stefik on July 6, 2001 and one by Barbara Stefik on August 27, 2001.

12. Or Ockham's Razor. The simplicity principle is credited to William of Ockham.

13. There are three main levels of formal languages for describing computation in computer science. In order of decreasing power, these are *context-sensitive*, *context-free*, and *finite-state* (or *regular*). The most powerful languages—the context-sensitive languages—require more computer memory and a more complex system to recognize correct expressions. Finite-state languages are much

less powerful. The powerful formalisms can do the work of the simpler ones, but not the other way around. It is not possible in general to write a finite-state program that will correctly recognize a context-sensitive language.

14. In the language of algorithmic analysis, the programs had greater "time complexity" than Kaplan expected.

15. Kaplan's reference to a pony refers to an old joke about two boys getting presents for Christmas. One boy was an optimist and the other a pessimist. In the story, the parents are testing the children and give the pessimist the most wonderful collection of gifts. He stays true to character and complains that the toys will probably just break. They give the optimist son a pile of manure in his room. He opens the door and says "Oh boy! There has to be a pony in here somewhere!"

16. Finite-state grammars can be used to create "finite-state machines." These "machines" are often represented as circle and arrow diagrams where the circles represent states and the arrows have labels representing conditions for going to the next state. The grammars and the machine diagrams are equivalent and it is easy to translate from one format to the other.

17. For readers familiar with predicate calculus notation, Kaplan's "not there exists not" observation refers to the logical equivalence $\forall x\ F(x) \equiv \neg(\exists x\ \neg F(x))$. In plain English, Kaplan's insight amounts to noticing that "There are no apples that are not tasty" means the same thing as "All apples are tasty." The importance of this insight was in the transformations of representation that it made possible. The context-sensitive style rules were easy to understand, so it was easy to determine that they were correct. Unfortunately, a full context-sensitive morphology computation was very expensive. Kaplan realized that a finite-state representation could do the same work much more efficiently. The finite-state rules, however, were bewildering to understand. Kaplan's "Aha!" led to two key insights. The first was that the full power of the context-sensitive rules was not used, so finite-state rules could do all the work that was really needed. The second was that an engineering regime and compiler technology could automatically translate the elegant context-sensitive rules into finite-state rules that were very efficient.

18. For this story, Mark Stefik was interviewed by Barbara Stefik in July 2002.

19. At the time of this writing, the digital publishing industry is a long way from adopting digital rights and using it in a way that satisfies consumers. Sales of compact discs for music have been dropping for a couple of years, and various copy-protection schemes have not met consumers' expectations. At the same time, some of the larger commercial interests are making substantial bets that digital rights management can be made attractive for the consumer market.

20. PCMCIA cards are used especially with laptop computers for adding functionality, such as modems, wireless modems, and extra disks.

21. See Csikszentmihalyi, *Creativity*.

Chapter 7

1. See Richard Rosenbloom, "Sustaining American Innovation: Where Will Technology Come From," in National Research Council, *National Forum on Science and Technology Goals*.

2. Ted Selker's story is based on an interview recorded April 18, 2001.

3. Anti-lock brakes for cars are designed to modulate brake pressure on slippery roads and do other things to give a safer response.

4. Dave Fork's story is based on an interview recorded August 25, 2002.

5. CMOS (complementary metal oxide semiconductor) is the preferred silicon technology for low energy consumption.

6. Pattie Maes's story is based on two interviews, one conducted by Mark Stefik recorded June 28, 2001 and one by Barbara Stefik on September 7, 2001.

7. Cited in *Sparks of Genius*, p. 250.

8. Sheridon's comments here seem to refer to a phase of invention in which ideas are "cooking" in the unconscious, rather than being about the entire experience of invention. In other parts of the interviews, Sheridon refers to social experiences inspiring and facilitating invention as a member of a research group.

9. Quoted from *Sparks of Genius*, p. 4.

10. Quoted from pp. 184–185 of Horace Judson's book *The Search for Solutions*.

11. Diana Smetters's story is based on an interview recorded April 12, 2001.

12. Based on a personal communication dated April 19, 2001.

13. Barbara Stefik's story is based on an interview by Mark Stefik recorded August 16, 2002.

14. From an interview by Kristi Helm published in the *San Jose Mercury News* under the headline "Redefining the PC" (January 11, 2003).

15. Source: personal communication dated March 10, 2003.

16. In security research, building a defense against an attack requires making estimates about the capabilities of adversaries. A lone hacker as an adversary is generally considered to be less threatening than a well-funded government agency with many theoreticians and extremely fast computers.

17. Mark Stefik's story was recorded in an interview conducted August 16, 2002.

18. More precisely, Dendral was the name of the first program in a series of structure elucidation systems. Dendral worked on tree-shaped chemical structures, that is, structures without cycles. It was also the name of the project. Congen, a program developed later in the project, was capable of handling cyclical structures.

19. Djerassi himself reports this event in his autobiography, *The Pill, Pygmy Chimps, and Degas' Horse*.

20. See J. Lederberg, "Digital Communications and the Conduct of Science: The New Literacy," *Proceedings of the IEEE* 66 (1978), no. 11: 1314–1319. The article is excerpted in Stefik, *Internet Dreams*.

21. Agar is a nutrient gel that is placed in Petri dishes and used as a medium for growing cultures of bacteria.

22. Cited on p. 247 of *Sparks of Genius*.

Chapter 8

1. Statistical image analysis is concerned with analyzing the content of images. Baird's work has focused largely on recognition of text. The field is statistical in that it has formal statistical models of error—accounting for things like skew, noise, and so on.

2. Henry Baird's story is based on two interviews, one conducted by Mark Stefik on March 15, 2001 and one by Barbara Stefik on June 27, 2001.

3. CAD: computer-aided design.

4. Card's story is based on two interviews, one conducted by Mark Stefik on July 6, 2001 and one by Barbara Stefik on August 2, 2001.

5. Card is using the term "move" (analogous to a move in chess) to refer to an operation in a combat situation. For example, it might take 5 seconds to pick a target and aim.

6. Not all chess matches (particularly those between Grand Masters) are played with such rapid pacing.

7. Chunking—the grouping of several concepts into a single "chunk" for the purpose of reasoning about it as a whole—is crucial for the learning that speeds up performance in repeated tasks.

8. Protocol analysis is a method of studying someone carrying out a task by encoding his actions in a "protocol" of steps and then analyzing the log of steps for meaningful patterns.

9. Acoustic holography is a method of recording and displaying images using sound waves.

10. Sheridon's story is based on two recorded interviews, one conducted by Mark Stefik on March 17, 2001 and one by Barbara Stefik on June 12, 2001.

Chapter 9

1. Bob Taylor's story is based on an interview recorded July 3, 2001.

2. "Dealer" is the name of a weekly laboratory meeting of the Computer Science Laboratory at PARC. At a Dealer, a speaker gives a talk and determines at the beginning how the group will handle questions. That is "Dealer's choice."

3. Joshua Lederberg's observations are drawn from an interview recorded August 26, 2002.

4. Springtime is sometimes called "lambing season," meaning that this is when PhDs graduate and become available for hiring.

5. John Hennessy's thoughts are drawn from an interview recorded October 10, 2002.

6. Henry Baird's story is drawn from interviews recorded March 15, 2001 and June 27, 2001.

7. David Fork's story is drawn from an interview recorded August 25, 2002.

8. A "bootleg" project is one that is running on the remainders of the resources of an official project. Managers and researchers tend to keep these projects under the radar, protecting them from too much attention or scrutiny until they are strong enough. The term "bootstrapped project" is also used, suggesting the "pulling yourself up by your bootstraps" metaphor.

9. EML: the Electronic Materials Lab at PARC.

10. The inventory doesn't pick up everything; some projects are new and emerging, and some bootlegged projects are consciously excluded from premature review. Researchers at PARC have always had the latitude to start exploratory projects without much management oversight. This practice dates to PARC's beginnings, when George Pake (PARC's founder) set up the initial culture. In later writings about PARC, Pake said "Little success is likely to come from showing researchers to a laboratory, describing in detail a desired technology or process not now existent, and commanding: 'Invent!' The enterprise will go better if some overall goals or needs are generally described and understood and if proposals for research are solicited from the creative professionals. Managing the research then consists of adjusting the budgets fore the programs to give selective encouragement." Pake was revered at PARC, not only for founding the establishment and protecting it through different corporate struggles and changes of administration, but also for knowing how much to "stay out of the way" of creativity.

11. Smalltalk is a programming language and environment developed at Xerox PARC in the late 1970s. It was the first practical object-oriented programming system, and as such became a major influence on many programming languages that followed. A classic book on Smalltalk is Adele Goldberg and Dave Robson, *Smalltalk 80: The Language* (Addison-Wesley, 1989).

12. Dave Robson's story is based on a conversation with Mark Stefik on April 2, 2003.

13. "Dynabook" was the name of an Alan Kay "dream machine"—an early sketch of what he thought a personal computer should be like. The Dynabook that Kay sketched looks a lot like a present-day laptop computer.

14. Bitblt (pronounced "bit-blit") is a procedure for moving a rectangular region of bits on the from one place on the display to another.

15. Ron Kaplan's observations are drawn from interviews recorded July 6 and August 27, 2001.

16. NLLT: the Natural Language Theory and Technology group at PARC.

17. Mark Stefik's observations are drawn from an interview conducted by Barbara Stefik in August 2002.

18. Dealer meetings still happen once a week at CSL.

19. Daniel Bobrow's observations are drawn from an interview recorded September 3, 2002.

20. Newell's quote is drawn from an interview of Stu Card conducted July 6, 2001.

21. This story is based on a personal communication from Stu Card dated March 29, 2003.

Chapter 10

1. This quotation was a source of much debate between the authors of this book. From one perspective, the "unreasonable man" seems to be completely out of touch with the environment. He may be the source of so-called progress, but at what expense to the environment? From another perspective, the "unreasonable man" may be responding more to the social world, anticipating what is ahead. He may in fact be completely in tune with an inner guide but far ahead of the social norms of his time.

2. Breakthrough inventions and their obstacles relate to the question "What is possible?" When there are serious breakthrough obstacles, answering "What is possible?" is very difficult and a breakthrough is needed.

3. Radical innovations and their obstacles relate to the question "What is needed?" When an invention is aimed at future customers and their future needs, answering "What is needed?" is very difficult. An invention can be ahead of its time for several reasons. For example, the need may be just emerging, the invention may need to evolve substantially to meet the need more perfectly, or there may be many elements of co-inventions or infrastructure that are needed before the invention can be practical.

4. Technically, packet switching made *ARPANET* possible. ARPANET then evolved into the Internet. Several well-written histories of the ARPANET and Internet are available online at the website for the Internet society. See www.isoc.org/internet/history/.

5. Packet switching solves several problems of data communication. It is robust, since the nodes can re-direct information along any available route if nodes are busy or broken. It also provides a means for sending very large amounts of information without requiring all the intermediate nodes to have enough local memory to store the whole message.

6. Kahn's story is based on an interview recorded September 23, 2002.

7. Taylor's story is based on an interview recorded July 3, 2001. In 2004, Robert Taylor, Charles Thacker, Alan Kay, and Butler Lampson received the Charles Stark Draper prize from the National Academy of Engineering for the invention of the first networked personal computer. This is the NAE's highest honor. They

cite it as a technological achievement that has changed almost every aspect of our lives. William Wulf, president of the NAE, said "these four prize recipients were the indispensable core of an amazing group of engineering minds that redefined the nature and purpose of computing." Reflecting on the award on National Public Radio's *Talk of the Nation*, Bob Taylor remarked that the personal computer was a collective invention, invented by the "community" at PARC.

8. Bravo was a visual text editor developed at PARC for the Alto computer. The creator of Bravo, Charles Simonyi, later worked for Microsoft, where he is known as the "father of Word." The similarities between the two programs run deep. For example, even the command names are the same, including the somewhat obscure single-character abbreviations of <control>-x, <control>-c, and <control>-v for the Cut, Copy, and Paste commands.

9. SmallTalk was one of the first and most influential of the object-oriented programming systems. It extended ideas from earlier systems like Simula-67, and it embedded them in a programming system with a graphical user interface. SmallTalk was developed at PARC.

10. At the low volumes of manufacturing for new technology, personal computers were also quite expensive. The Star workstation in 1981 was priced at $16,595, about five times the price of other personal computers on the market. Another factor was that the Xerox sales organization, accustomed to selling copiers, did not understand how to sell computer systems.

11. At that time, wiring a local computer network was beyond the capabilities of a reasonably handy individual, and contractors were not familiar with the technology or its requirements.

12. See Christensen, *The Innovator's Dilemma*.

13. Ted Selker's story is based on an interview conducted April 18, 2001.

14. At the time of this writing, Xerox and the companies involved in manufacturing the Palm Pilot are negotiating about royalties relating to the contributions of Unistrokes to Graffiti.

15. Liveboards were computer-based electronic whiteboards for use in meetings. PARC spun off a company to manufacture and sell liveboards in the late 1990s.

16. David Goldberg's story is based on two interviews, one conducted by Mark Stefik on March 29, 2001 and one by Barbara Stefik on April 2, 2001.

17. This issue of "clicking sounds" may seem quaint today for two reasons: laptops are much more common today than they were then, and keyboards are much quieter.

18. The space was about an inch square, roughly the size of a postage stamp.

19. Dave Patterson is a professor at the University of California at Berkeley. Along with John Hennessy, he developed RISC (reduced instruction set computers). He has also been a program manager at DARPA.

20. At PARC, an IP (invention proposal) is a formal document that an inventor creates in order to start a patenting process. It is a description of an invention and is signed and witnessed.

21. At PARC, a TAP (technical advisory) panel, reviews invention proposals and makes recommendations about patenting. It offers suggestions about priorities, related inventions, and sometimes further developments of ideas.

22. Mark Weiser was a Xerox PARC inventor and the manager of the Computer Science Laboratory. Among the things for which he is remembered is his leadership in establishing research on ubiquitous computing.

23. CHI: a conference on Computer and Human Interaction.

24. Tom Webster was a PARC patent attorney.

25. Donald Norman, a cognitive psychologist, was then at the University of California at San Diego. He later became a research fellow at Apple, then an executive at Hewlett-Packard. He is currently a professor at Northwestern University and a member of the Nielsen-Norman Group, a consulting organization concerned with usability and user interface design. He is well known for his 1988 book *The Psychology of Everyday Things*.

26. Bill Buxton is one of the influential researchers in human-computer interaction. At that time he was a professor at the University of Toronto.

27. Mark Stefik's story is based on an interview conducted by Barbara Stefik in July 2002.

28. This paper, "The Bit and the Pendulum," was reprinted in Stefik, *The Internet Edge*.

29. The paper, "Letting Loose the Light," appears in Stefik, *Internet Dreams*.

30. At the time of this writing, a major law suit by InterTrust against Microsoft was pending, with claims against a broad range of Microsoft technology integrated around trusted systems.

31. Xerox had a long history of working with publishers on copyright issues. It also had ambitions of providing document repository services to publishers. DRM would have been a better fit for Xerox if those businesses already had plans or understanding of digital publishing.

32. The research on small devices fitted in with PARC's early investments in ubiquitous computing and preceded the successful marketing of the Palm Pilot and PDAs by other companies.

33. For a variety of reasons, the original guerilla team did not join Content-Guard. Bettie Steiger retired from Xerox. Prasad Ram formed his own company, Savantech. Stefik had already returned to research, to work on sense-making technology.

34. In 2002 the National Science Foundation had a budget of about $3 billion for R&D across the sciences, DARPA had an R&D budget of about $2.3 billion, and NIH had a budget of about $22 billion for R&D in the health sciences. Figures are based on the NSF Survey of Federal Funds for R&D. According to the analysis in *IEEE Spectrum* (September 2002), the federal government funds about half of the basic research in the U.S. U.S. industry spent about $181 billion on R&D in 2000, about 71 percent of it on development.

35. Jerry Hobbs's observations are drawn from an interview recorded August 28, 2002.

36. People commonly leave words out of sentences where the meaning is clear to a listener. In this example, a verb phrase is left out. A more complete and logical version of the sentence would be "John called his mother and Bill called his mother too."

37. In information retrieval, precision measures the fraction of the documents retrieved that were in the desired set. Recall measures the fraction of the desired set that were retrieved. In the MUC competition, these measures were used for information-extraction tasks such as measuring whether certain facts or named objects were recognized and called out by the natural-language systems.

38. Today many American college students have only mobile phones. In countries where telephones were introduced only recently, phone cables are not being laid.

39. Subvocalizing: moving the tongue and the lips almost imperceptibly, creating a sound without "breathing out" or using the vocal cords.

40. As was mentioned earlier, the Internet was originally called ARPANET.

41. In the mathematics of percolation systems and constraint propagation systems, such spreading phenomena arise when the degree of interconnectedness or fan-out among the interacting elements is greater than a critical value or "threshold." The degree of interconnectedness governs the amplification of a phenomenon in a network because it represents how effects can fan out from one item to the next. When the fan-out is too low, spreading is unlikely to happen. When it is high, spreading happens quickly. In many systems, the threshold is discrete and is an important property of the network.

Chapter 11

1. Gibbons is a former Dean of Engineering at Stanford University.

2. For a detailed history of a later period in RCA history during which the development of the VideoDisc occurred, see Graham, *The Business of Research*.

3. Baird's accounts of innovation cultures from when he worked at Sarnoff Labs and Bell Labs are drawn from interviews recorded March 15, 2001 and June 27, 2001, from several personal communications, and from an informal talk he gave at PARC.

4. Jerry Hobbs's story is based on an interview recorded August 28, 2002.

5. In 1968, Doug Engelbart and his team from SRI's Augmentation Research Center staged a 90-minute public demonstration at the Fall Joint Computer Conference in San Francisco. The demonstration showed a computer system with multiple windows controlled by a mouse and used for hypertext linking, real-time text editing, and video teleconferencing. In 2000, Engelbart was awarded the National Medal of Technology.

6. From a conversation at PARC in November 2002.

7. Stewart Brand's book *The Media Lab* contains a very readable description of the institution shortly after its founding.

8. Pattie Maes's story is based on interviews recorded June 28, 2001 and September 7, 2001.

9. See Gordon E. Moore, "Some Personal Perspectives on Research in the Semiconductor Industry," in *Engines of Innovation*, ed. Rosenbloom and Spencer.

10. Robert Noyce, one of the inventors of the integrated circuit, was a founder of Fairchild Semiconductor and later (with Moore) a founder of Intel.

11. Reported in the *New York Times* on December 26, 2002.

12. According to an article in the *Seattle Times* (December 2, 2002), Bell Labs shrank from 30,000 employees in 1999 to 10,000 employees in 2002. The number of scientists doing core research shrank in the same period from 1,000 to 500.

13. According to an article in the *Seattle Times* (December 2, 2002), Lucent lost $11.9 billion in 2002 and cut 30,000 jobs. In the March 2003 *Scientific American*, Gary Stix reported several other statistics about the organization in his article "The Relentless Storm." Bell Labs Research cut its workforce from 24,000 employees in 1999 to 10,000 and shut down its Silicon Valley research center. It reduced its research and investment spending from $3.54 billion in fiscal 1999 to $2.31 billion in fiscal 2002. After the labs were acquired by Lucent, the microelectronic, fiber, and business-networking divisions were jettisoned—weakening the rationale for a physical sciences unit.

14. Dan Russell's story is based on several personal communications and on a public talk he gave on this topic at a PARC forum on June 13, 2002.

15. DB2 is a major IBM database product. Lotus makes office information systems, notably Lotus Notes.

16. In the early 1990s, Apple came out with QuickTime, the first widely available software player for digital movies. In the first versions, the windows were small and the display of the movies was jumpy.

17. Variants of Moore's Law predict that the memory on a chip can double every 18 months. Historically, Moore's Law has held pretty well and has been based primarily on processing improvements enabling chip manufacturers to make smaller and smaller (faster and faster) circuit elements on the chips.

18. For a discussion of these intermediate terms in the context of an approach to innovation that is open to both internal and external technology as well as existing and new markets, see Chesbrough, *Open Innovation*.

19. On p. 3 of *The Business of Research*, Margaret Graham writes: "In any case, if a proposed innovation is really new, little information about the market can be trusted. Often other factors, such as the behavior of key competitors, or the predictions of the press, provide the only information available."

20. George Heilmeier developed his catechism when he was the director of DARPA, where he managed projects having to do with stealth aircraft, lasers, and artificial intelligence. Heilmeier is known for his invention of an early version

of liquid crystal displays using the dynamic scattering method. He is currently Chairman Emeritus of Telcordia Technologies. The catechism evolved for many years, and there are now several variants of it. This version is representative.

21. This quotation is from an interview with Chuck Thacker recorded September 6, 2002.

Chapter 12

1. The model presented here is simplified for pedagogical purposes. The pictorial analogy of jumping over hurdles overlooks the collaborative nature of invention, and the workings of collective action and synergy. It does not take into account collaborative action in which one person takes a solution part of the way and a second person takes it further. In terms of jumping over hurdles, it does not model how one person can give another jumper a "boost" to get over a barrier. Various theoretical models predict how it is that groups can perform better than individuals. Small groups which include people of extremely high performance in different areas can be particularly effective. Many of the best research centers deliberately hire "good nervous systems" so that high-performance multi-disciplinary groups can form spontaneously and work together.

2. We could extend the analogy to reflect collaborative effort. In the case of basic research, we could (at least in principle) postulate that scientists from multiple fields work together. However, institutions for basic research do not tend to include people who would have connections or expertise in understanding needs. This limits the potential effectiveness for collaborating in a way that would effectively address both "What is possible?" and "What is needed?" The same issue arises for groups at the product-development end. Again, a plus of corporate research organizations is that they often have cultures that deliberately bring together people with these different kinds of expertise to search out key problems.

3. See Chesbrough, *Open Innovation*.

4. Kahn's comments are drawn from an interview recorded September 23, 2002.

5. The last point about holding intellectual property tightly did not usually work out this way in typical practice. Most large companies in an industry develop a patent portfolio, and then find themselves facing interlocking legal battles with competitors with their own patents. Faced with the uncertain and costly option of trying to enforce their patents, the most common practice has been for large companies to enter into cross-licensing agreements. The effect of this has been to inhibit the entry of new competition, since the network of interlocking patent agreements between the major players provided a formidable barrier to entry.

6. The inventors of the personal computer at PARC realized this quite well. The business logic of Xerox and the $17,000 price tag, however, made marketing the computer to individuals quite impractical.

7. IBM tried to regain a proprietary advantage by introducing OS/2, a proprietary operating system intended to compete with Microsoft. However, OS/2 came

along too late. Supported by a single processor at the base, the computing industry was already in transition from a vertically integrated industry in which companies supplied everything to their customers to a horizontally integrated
industry in which companies competed in the own layers: processors, computers, operating systems, applications. Competition brought prices down and inhib-
ited a return to the old style.

8. John Seely Brown's observations are from an interview recorded September
17, 2002.

9. An example of a new kind of market was "Webvan," an Internet-based service
for ordering groceries and having them delivered to your home. Webvan was
very popular among people who didn't want to spend time shopping, but it tried
to expand too fast and it lacked a sustainable revenue model.

10. In other words, is there really a market for this product? Another variant
on the "dog food" metaphor is a slogan used by developers developing new technology—"Eat your own dog food" or even "dogfooding" for short. This means
that the creators directly use the new technology in their own lives, thereby
encountering its limitations firsthand and being inspired to improve it.

11. For a description of the culture and the steps in founding a company using
venture-capital funding, see Kaplan, *Startup*.

12. Bettie Steiger's observations were drawn from an interview recorded
September 20, 2002. In *The Innovator's Solution*, Christensen and Raynor also
reflect on the rise and fall of the dotcom bubble. One of the themes of their book
is that corporations are expected to grow, and that when they get large a 10%
growth requires a lot of revenue. Before the dotcom cycle, venture capitalists had
a reputation for funding startups more effectively than corporations. Why would
people with such experience invest so much money is startups that had no customers or products? The answer is that investors had put massive amounts of
capital into their funds, expecting to earn the historically high rates of return.
As the funds grew, they could not sustain growth making the $2 million and $3
million early-stage investments of the past and did not have enough partners or
good options to do that. Like corporations, they had become big and had to
make big investments that would grow fast. In effect, they had moved up market
into the sizes of investments that are usually handled at later stages.

13. By January 2003, returns on venture-capital investments reached the worst
slump in three decades. As reported in an article by Matt Marshall in the January
15, 2003 *San Jose Mercury News,* the average annual returns had dropped to
14.5%. Returns from venture-capital firms had dropped by an average of 22.3%
for the year ending in September 30, 2000. In contrast, some of the best firms
had 50% returns annually over 5 years or more in the mid 1990s. Returns at
the three-year mark for venture firms, 420% in 1996, were down to 53% by
1998 and down to negative 32% in 1999.

14. UNIX developed a reputation as a simple but solid operating system. Bugs
in new releases were quickly addressed and fixed.

15. The four "NBIC" areas are nanotechnology (very-small-scale materials
and devices), biotechnology (including gene and protein analysis), information

technology, and cognitive science. See e.g. Roco and Bainbridge, *Converging Technologies for Improving Human Performance.*

16. Paul Saffo's story is drawn from an interview recorded October 10, 2002.

17. MEMS: MicroElectricalMechanical Systems. Much of the MEMS research is funded by DARPA with the goal of developing advanced technology that combines sensing, actuating, and computing. Current fabrication for MEMS grows out of the technologies used for semiconductor devices. MEMS projects funded by DARPA are in areas of fluid sensing and control, mass data storage, optics and imaging, inertial instruments, radio-frequency components, and sensor and actuator arrays and networks.

18. Photovoltaics are technologies for making photocells.

19. John Hennessy's observations are drawn from an interview recorded October 10, 2002.

20. The Clark Center, a new research institute at Stanford officially dedicated in October 2003, is at the core of a broad-based multi-disciplinary initiative at Stanford called Bio-X. Bio-X will bring together researchers from the biosciences, physical sciences, and engineering to focus on multi-disciplinary biomedical research. The Clark Center is funded largely by a gift from James Clark, a former Stanford professor and a founder of Silicon Graphics and Netscape. Symbolically, the Clark Center is located between Stanford's hospital and its West Campus of science and engineering buildings. There are about 25 academic departments represented in the center.

21. Joshua Lederberg's story is drawn from an interview recorded August 26, 2002.

22. David Fork's comments are from an interview recorded August 25, 2002.

23. The term "Goldilocks zone" derives from the children's tale "Goldilocks and the Three Bears." In the story, Goldilocks tries three bowls of cereal; one bowl is too hot, one is too cold, and one is "just right." In a Goldilocks zone of risk, the penalty/reward structure is "just right" to encourage innovation.

24. "Intrapreneurship" refers to programs to encourage innovation and spinoffs to occur from within large corporations.

Bibliography

Publications

Allen, Thomas J. *Managing the Flow of Technology: Technology Transfer and the Dissemination of Technological Information within the R&D Organization.* MIT Press, 1977 and 1984.
How research and technology transfer work in large organizations.

Bennis, Warren, and Patricia Ward Biederman. *Organizing Genius.* Perseus Books, 1997.
A collection of stories about high-functioning teams.

Berliner, Paul F. *Thinking in Jazz: The Infinite Art of Improvisation.* University of Chicago Press, 1994.
An evocative and insightful book based on interviews with jazz musicians. Although not about high technology, it is about the direct experience of improvisation, the rich process of preparing the mind for creative performances, and mentoring and apprenticeship.

Blonder, Greg. "Q&A: The Innovation Machine Needs Fixing." BusinessWeek online, August 5, 2002.
A short piece arguing that America's innovation engine is broken and has become stuck in a climate of short-term thinking.

Bok, Derek Curtis. *Universities in the Marketplace: The Commercialization of Higher Education.* Princeton University Press, 2003.
A look at the conflicts of interest that arise when universities become involved in commercialization of their research. A fundamental message is that having goals beyond those of the educational mission can undermine trust between faculty and students.

Brand, Stewart. *The Media Lab: Inventing the Future at MIT.* Viking Penguin, 1987.
Stories about the people, their methods, and their vision at MIT's Media Lab shortly after its founding.

Brown, John Seely. *Seeing Differently: Insights on Innovation.* Harvard Business School Press, 1997.

An edited volume of articles from the *Harvard Business Review* exploring the "seeing differently" metaphor and arguing that new strategies for continuous innovation in business require constant reexamination of assumptions and new ways of gathering information.

Brown, John Seely. "Seeing Differently: A Role for Pioneering Research." *Technology Management* 41 (1998), no. 3: 24–33.
How breakthrough or pioneering research explores the "white space" between disciplines and creates options for the future.

Chesbrough, Henry. *Open Innovation: The New Imperative for Creating and Profiting from Technology.* Harvard Business School Press, 2003.
A very readable book, based on case studies, that explains the logic and history of "closed innovation" and why it is no longer the right kind of strategy.

Christensen, Clayton. *The Innovator's Dilemma.* Harper, 2000.
Case studies of why companies focus their attention and priorities on short-term development of current products and resist or squelch disruptive innovation and why this hurts them in the long term.

Christensen, Clayton, and Michael Raynor. *The Innovator's Solution.* Harvard Business School Press, 2003.
A follow-up to Christensen's earlier work, this book addresses how the imperative for growth leads businesses towards strategies that maxmize present profits at the expense of future growth. This book is an accounting of the business logic of innovation. It provides strategic guidance for identifying customers and kinds of markets in which growth is possible, and also an analysis of strategies for sustaining growth and profitability when products become commodities.

Csikszentmihalyi, Mihaly. *Creativity: Flow and the Psychology of Discovery and Invention.* Harper Collins, 1996.
A wide-ranging book based on interviews of 91 creative individuals.

Drucker, Peter F. *Innovation and Entrepreneurship.* Harper Business, 1985.
Drucker has been an influential leader in business thinking for many years. This book is a classic on innovation, focusing on improving organizational practices.

Engardio, Pete, Aaron Bernstein, and Manjeet Kripalani. "The New Global Job Shift." *BusinessWeek*, February 3, 2003.
A comprehensive collection of articles about globalization and its impact on U.S. companies and U.S. jobs.

Feder, Stanley A. "Overcoming 'Mindsets': What Corporations Can Learn from Government Intelligence Failures." *Competitive Intelligence Review* 11 (2000), no. 3: 28–36.
Intelligence failures in the United States and other countries have been a subject of retrospective analysis for decades. Feder believes that "mindsets" are the major source for government intelligence analysis. A mindset is a "fixed attitude or state of mind that provides a (sometimes wrong) context for the interpretation of data or events or for making decisions."

Feynman, Richard P. *The Pleasure of Finding Things Out*. Helix/Perseus, 1999.
A famous physicist discusses his experiences growing up, as a graduate student, and as a teacher and a scientist. The emphasis is on the playful side of creativity.

Goldstein, Harry. "They Might Be Giants." *IEEE Spectrum*, September 2002: 43–48.
Every year, *IEEE Spectrum* (a journal of the Institute of Electrical and Electronics Engineers) publishes a special report summarizing the state of the R&D. This one includes tables on R&D spending by the 100 largest corporations and descriptions of emerging and leading-edge technologies.

Gomory, Ralph E., and Roland W. Schmitt. "Science and Product." *Science* 240 (1988): 1131–1204.
A widely circulated article about what is needed beyond "science" in order to translate leadership in scientific discoveries into industrial or product leadership, written at a time when Japanese dominance in electronics was first being felt in the U.S. electronics industry (which had historically spent more on R&D). It emphasizes several lessons in product-development time, design for manufacturability, and sustained incremental improvement.

Graham, Margaret B. W. "Industrial Research in the Age of Big Science." *Research on Technological Innovation, Management, and Policy* 2 (1985): 47–79.
A historian's perspective on the evolution of industrial research organizations in the 20th century.

Graham, Margaret B. W. *The Business of Research: RCA and the VideoDisc*. Cambridge University Press, 1986.
From the editor's preface: "RCA, as Graham shows, was for many decades a remarkably successful, high-tech business. Under the leadership of David Sarnoff, the relationship between the R&D organizations and the rest of the corporation was managed with great skill. When for a variety of reason that Graham describes, the managerial task was no longer performed successfully, the firm experienced serious problems translating sophisticated technological concepts into profitable products." This monograph, written over a 10-year period, was based on extensive interviews and documentation from RCA.

Graham, Margaret B. W., and Bettye H. Pruitt. *R&D for Industry: A Century of Technical Innovation at Alcoa*. Cambridge University Press, 1990.
This in-depth historical perspective on the evolution of industrial research in the aluminum industry from 1888 to 1983 shows how periods of mismatch occurred in the cultures of innovation as the businesses matured and as international competition became significant.

Graham, Margaret B. W., and Alec T. Shuldiner. *Corning and the Craft of Innovation*. Oxford University Press, 2001.
A detailed account of innovation in a technology company that has sustained investment in R&D for 150 years.

Grove, Andrew S. *Only the Paranoid Survive: How to Exploit the Crisis Points That Challenge Every Company*. Bantam Doubleday Dell, 1996.
A very readable book, by one of the founders of Intel, on the subject of responding when the basics of a business change drastically. Grove cites six forces that affect the well-being of a business.

Hadamard, Jacques. *The Psychology of Invention in the Mathematical Field*. Dover, 1945.
An exploration of the thought processes of several creative mathematicians and scientists, including Dalton, Pascal, Descartes, and Einstein.

Hardy, G. H. *A Mathematician's Apology*. Cambridge University Press, 1940.
An account of a life in mathematics as the life of a creative artist.

Hiltzik, Michael. *Dealers of Lightning: Xerox PARC and the Dawn Of The Computer Age*. Harper Business, 1999.
The story of the invention and the inventors of the personal computer at PARC in 1973.

Judson, Horace Freeland. *The Search for Solutions*. Holt, Rinehart, and Winston, 1980.
This well-illustrated book for young readers on what doing science is like contains many quotations from interviews of scientists.

Kaplan, Jerry. *Startup: A Silicon Valley Adventure*. Penguin, 1994 and 2002.
This fast-paced book on the founding of a company in Silicon Valley goes through the steps of attracting venture capital funding, assembling a management team, hiring a staff, refining products, surviving multiple rounds of funding, and so on.

Kelley, Tom, with Jonathan Littman. *The Art of Innovation. Lessons in Creativity from IDEO, America's Leading Design Firm*. Doubleday, 2000.
The story of how innovation works at IDEO.

Kohn, Alfie. *No Contest. The Case Against Competition*. Houghton Mifflin, 1986 and 1992.
How the competitive reward structures put in place for teams influences their ability to be creative and innovate.

Lakatos, Imre *Proofs and Refutations: The Logic of Mathematical Discovery*. Cambridge University Press, 1976.
An exploration of how an attempt to prove theorems leads to insights into how they are wrong and what the more subtle mathematical truths might be.

Moore, Geoffrey A. *Crossing the Chasm: Marketing and Selling Technology Products to Mainstream Customers*. HarperBusiness, 1991.
This exposition of business strategies and the marketing of technology contains a particularly accessible characterization of the technology-adoption curve.

Morita, Akio. *Made in Japan: Akio Morita and Sony*. Penguin, 1988.
The story of the founding, struggles, and amazing success of the Sony Corporation.

National Commission on Excellence in Education. *A Nation at Risk: The Imperative for Educational Reform.* 1983.
A special commission's report on education in the United States.

National Research Council. *National Collaboratories: Applying Information Technology for Scientific Research.* National Academy Press, 1993.
A report on several experiments in linking researchers and institutions with information technology to enable sharing and remote use of research equipment.

National Science Board. *Science and Engineering Indicators—2002.* National Science Foundation, 2002.
The fifteenth such report prepared since 1950, this provides a broad base of quantitative information about U.S. science, engineering, and technology for use by the public and private policy makers. It analyzes the scope, quality, and vitality of the U.S. science and engineering enterprise.

Platt, John Rader *The Excitement of Science.* Houghton Mifflin, 1962.
A wide-ranging analysis of the methods and challenges of the most productive scientists, their social values, and how creativity is motivated. The chapters on the working styles of different scientists are particularly interesting.

Pretzer, William S., ed. *Working at Inventing: Thomas A. Edison and the Menlo Park Experience.* Johns Hopkins University Press, 1989 and 2001.
The history of Edison's "invention factory" between 1876 and 1882.

Roco, Mihail C., and William Sims Bainbridge. *Converging Technologies for Improving Human Performance: Nanotechnology, Biotechnology, Information Technology and Cognitive Science.* Kluwer, 2003.
Roco and Bainbridge argue that the technologies of the four previously separate fields mentioned in the title are converging. They discuss how this will affect working, learning, aging, group interaction, and human evolution.

Root-Bernstein, Robert, and Michèle Root-Bernstein. *Sparks of Genius: The Thirteen Thinking Tools of the World's Most Creative People.* Houghton Mifflin, 1999.
A delightful and readable account of how creative people work, with emphases on the use of imagination and the value of play.

Rosenbloom, Richard S., and William J. Spencer, eds. *Engines of Innovation: U.S. Industrial Research at the End of an Era.* Harvard Business School Press, 1996.
An edited collection of essays by technical managers from leading technical companies and scholars of history and economics exploring the implications of downsizing and redirecting of corporate research laboratories in the 1990s.

Saffo, Paul. "Untangling the Future." *Business 2.0,* June 2002.
A look at four major areas of rapid advances in technology: information technology, materials science, bioscience, and energy.

Smith, Douglas K., and Robert C. Alexander. *Fumbling the Future: How Xerox Invented, Then Ignored, The First Personal Computer.* Morrow, 1988.
A journalistic account of the business decisions and the people behind Xerox's decision not to develop the personal computer.

Stefik, Mark. *The Internet Edge: Social, Technical, and Legal Challenges for a Networked World*. MIT Press, 1999.
How the Internet created possibilities for change and greater connectivity when legal systems, social attitudes, economic systems, and technology were not yet ready for the changes.

Stefik, Mark, and John Seely Brown. "Toward Portable Ideas." In *Technological Support for Work Group Collaboration*, ed. M. Olson. Erlbaum, 1989.
An article on the use of whiteboards in public spaces at PARC, and on a proposal for creating electronic whiteboards to promote idea generation and persistent conversations in a research organization.

Stokes, Donald E. *Pasteur's Quadrant: Basic Science and Technological Innovation*. Brookings Institution, 1997.
How a linear model of basic research and applied research misunderstands the nature of research and distorts public policy.

Waldrop, M. Mitchell *The Dream Machine: J. C. R. Licklider and the Revolution That Made Computing Personal*. Viking, 2001.
An account of the thinking that led to the personal computer and to the idea that computers should help people both to think and to communicate.

Womack, James P., Daniel T. Jones, and Daniel Roos. *The Machine That Changed the World: The Story of Lean Production*. Harper, 1990.
In the 1990s there was a resurgence of interest in the re-engineering of business practices. This book, the product of a five-year study by MIT's International Motor Vehicle Program, analyzes a kind of breakthrough that is largely complementary to the stories in this book.

Websites

inventors.about.com

americanheritage.com/it/index.shtml

si.edu/lemelson

web.mit.edu/invent

usfirst.org/robotics/index.html

Index